Traveller's Guide
to the
Solar System

Giles Sparrow

Collins
An imprint of HarperCollins Publishers
77–85 Fulham Palace Road
London
W6 8JB

www.collins.co.uk

First published in 2006

11 12 10 09 08 07 06

10 9 8 7 6 5 4 3 2 1

A catalogue record for this book is available from the British Library.

ISBN-10: 0-00-723410-4
ISBN-13: 978-0-00-723410-3

Commissioned by Emily Pitcher
Edited by Caroline Taggart
Design and layout by Pikaia Imaging
Proofread by Dan Green

Colour origination by Dot Gradation, Essex
Printed and bound in Great Britain by CPI Bath

ITINERARY

Introduction

So you want to go into space? Well, it's little wonder. These days it seems you can't flick through the pages of a magazine or skip between holovision channels without being bombarded with pictures of exotic destinations. Whether it's a peace conference at the Presidential Estate on Asteroid 458390 Camp David, the Solar System Ringsurfing Championships in orbit above Saturn, or Hollywood starlets cavorting in the latest luxury orbital resort, everybody else seems to be space travelling, so why shouldn't you?

It's hard to believe, but in the first fifty years of the Space Age, between 1957 and 2007, fewer than 500 people became astronauts, and just a couple of dozen left Low Earth Orbit. These pioneers were rather like the explorers of an earlier age, who were funded by principalities and powers to seek out new territories and resources. Government-funded exploration opened the way for independent missions to follow, though at least the early astronauts didn't have to take beads and trinkets for the natives.

Admittedly the first super-rich independent space travellers were rather better off than the average gap-year student, but they still had to hitch lifts with someone who was going there anyway (usually someone Russian). Things began to change with the first commercial spacecraft – ingenious vessels that could offer the merely wealthy the chance of a few hours in orbit. As near-Earth space became more familiar, travellers followed quickly on the heels of government-funded explorers, first to the Moon, then Mars, and eventually across the entire Solar System.

There are package trips, of course, for those of you who like that sort of thing. If you want, you can book a weekend break on the Moon almost as easily as a flight to New Los Angeles or Santa Elvis. Launch site after breakfast , en route to the Moon by lunchtime, tea in lunar orbit, a night at the Tranquillity Ramada, a few hours at the Apollo 11 landing site, and back home for supper, laden with souvenirs and a complimentary photo of yourself next to the statue of Neil and Buzz.

But the *Traveller's Guide to the Solar System* is aimed far more at the independent tourist. We hope to recapture some of the spirit of adventure and mystery that accompanied the early decades of space exploration, and consequently you'll find that much of this guide is dedicated to the outer reaches and more exotic locales of the Solar System, as well as the more familiar and nearby destinations.

Of course, there are degrees of independence in space travel, so please don't think that this book is purely for the small minority of nuts-and-bolts survivalists. It's true that the outer reaches of the Solar System are still mostly the domain of real explorers – break down out there, and you're not going to be able to call out a repair truck. However, as far out as Saturn it's quite possible to sign up with small tour parties, perhaps tagging along on a commercial or scientific mission, or joining a dedicated tourist charter. Not only does this give you some back-up in case you run into trouble, it's also a great way to make new and lifelong friends.

Although we've now made our first reconnaissance of the Solar System, the dominion of the Sun is still full of new wonders to experience. We've only visited a fraction of all the worlds it has to offer, and this book can only touch on the largest and most popular or historically significant. Even professional explorers are lucky to see a handful of different worlds in their lifetimes, so choose your destination carefully – and enjoy!

 Good points:
Highlights of your destination

 Bad points:
Any drawbacks or dangers

 Day length:
Time for the planet to spin on its axis

 Year length:
Time the planet takes to orbit the Sun

 Gravity
Surface or cloudtop gravity – high, medium, or low

 Surface temperature

 Communications time
Time to receive a signal from Earth

 Danger!
Highlights specific dangers and risks

 The science bit
Theories and ideas

 History

 Travel Tips
Hints and tips for a better holiday

Travel essentials

As with any long journey, a trip across the Solar System will be a lot easier if you have some idea of where you're going and what you're likely to see on the way, and indeed if you know how to drive in the first place. This chapter gives you an introduction to our celestial neighbourhood and should tell you just enough about planetary motion and spacecraft navigation to avoid embarrassing blunders. It's also a good place to explain a little about spacesuits, which you'll be relying on a great deal, and to offer some general health advice.

Solar System basics

A holiday away from Earth isn't like your average package tour. While plenty of people visit Venice, the pyramids, and the ruins of Atlantis each year with barely the faintest idea of where they are on the map, space travel is more of a challenge. Assuming you don't want to limit your explorations to a rigorously regimented, scheduled lunar daytrip (and then why would you be reading this book?) you'll need some idea of Solar System geography.

Earth is the third of nine planets (some say eight, some say ten, but nine is kind of traditional), orbiting our local star, the Sun. The four worlds closest to the Sun – Mercury, Venus, Earth and Mars – are comparatively small and rocky (Earth is the biggest), and are collectively termed 'terrestrial planets'. Worlds five to eight are giants in comparison: Jupiter is the largest and innermost of these, followed in size and distance by Saturn. Uranus and Neptune are further still from the Sun, and smaller than the inner giants. Although all four outer worlds are enormous, they are made from much lighter elements than the terrestrial planets – hence the terms 'gas giant' and 'ice giant'.

Many of these eight planets are orbited by natural satellites or moons (Mercury and Venus are the only exceptions). Between and beyond the major worlds are countless smaller objects, sometimes lumped together as 'minor planets'. The majority are found in two bands – rocky asteroids concentrated between Mars and Jupiter, and frozen ice dwarfs beyond the orbit of Neptune. Pluto, the traditional ninth planet, is actually a large ice dwarf, but it's generally treated as a planet as well (for more on this, see Chapter 11).

Apart from the Sun, everything else in the Solar System is a moving target – and that includes Earth. All the planets move around the Sun according to a set of laws discovered by German astronomer Johannes Kepler in the early 1600s (see 'Planetary motions', opposite). The same laws also apply to moons in orbit around their planets.

GRAVITY BASICS

Every object in the Solar System has a gravitational field that attracts other objects. The strength of this field increases for objects with larger masses, but falls off rapidly as you get further from the centre of the field. Double an object's mass, and you double its gravitational force; double your distance from it, and you feel just one quarter of the force. The smallest objects with noticeable gravity are comet nuclei.

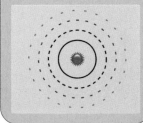

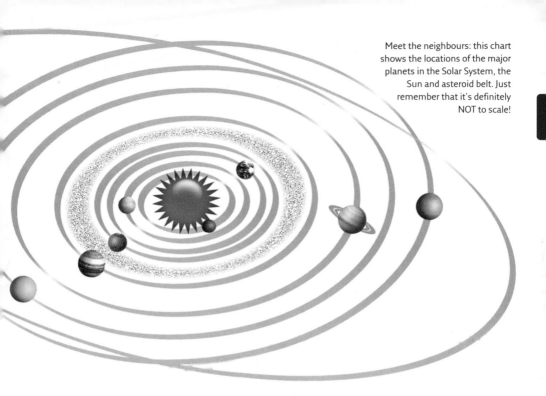

Meet the neighbours: this chart shows the locations of the major planets in the Solar System, the Sun and asteroid belt. Just remember that it's definitely NOT to scale!

Planetary motions

Put simply, Kepler's Laws explain that planets move around the Sun in ellipses (a sort of stretched circle with the Sun at one of two 'foci' or central points, as shown here). This means that the planet is closer to the Sun at some points in its orbit than at others: the closest point is called 'perihelion' and the furthest 'aphelion'. A circular orbit is just a special type of ellipse, with the two foci in the same place. The further a planet is from the Sun, the slower it moves along its orbit. So a planet moves faster at perihelion than aphelion (indicated here by the red arrows) and a planet in a closer orbit to the Sun moves faster than one further out.

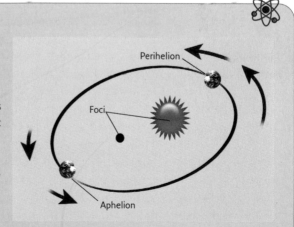

Perihelion

Foci

Aphelion

Most of the objects in the Solar System have orbits in or around a relatively flat plane, so that if you looked at them all side-on, they'd nearly line up. Most planets and asteroids are also in almost-circular orbits, and spin more-or-less 'upright' relative to their orbits (see p.206 for some explanations about all this).

For want of a better definition, the actual 'flat plane of the Solar System' is taken as the plane across the Earth's orbit. The technical term for this plane is the 'ecliptic' and any star atlas will show you how it runs across the sky, marking out the path that the Sun appears to take against the background stars (just don't go looking for a convenient dashed line on the sky itself). Of the other planets, the most significant 'inclinations' to this orbit are Mercury, tilted at 7°, and Pluto, tilted at 17°. So, tiny and distant Pluto aside, all the planets generally loiter close to the ecliptic and, seen from Earth, only ever appear against a narrow band of stars (the constellations of the zodiac).

MAGNETIC FIELDS

Most planets and several moons have a magnetic as well as a gravitational field around them. This makes them behave as if they have a giant magnet embedded inside them and is usually the result of molten metal swirling around somewhere inside them, either now or in their past. Magnetic fields affect magnetized or electrically charged objects, deflecting their movement – they emerge at a point on a planet's surface called the magnetic north pole, loop round the planet and plunge back in on the opposite side at the magnetic south pole. However, their effects are generally much weaker than those of gravity.

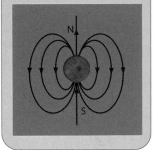

THE BIGGER PICTURE

Just in case a copy of this guide falls through a cosmic wormhole and into the hands of Space Aliens from Beyond the Stars (unlikely, but you never know), some directions might be useful. The Sun and its Solar System are located between two arms

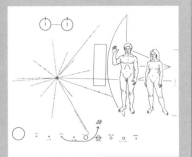

of the large spiral galaxy we call the Milky Way, within a small cluster of galaxies known, rather unoriginally, as the Local Group. We're about 26,000 light-years out from the old yellow stars at the galactic centre, but are also a comfortable distance from most of the violent star-forming regions in the spiral arms. For further details, see the maps freely distributed on a number of space probes in the late twentieth century.

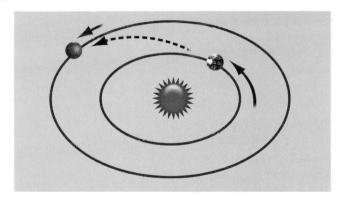

Flying from one planet to another isn't like flying in a straight line – for maximum fuel efficiency you'll want to follow a spiral path linking the two orbits and approach your target from behind. So it's important to plan the flight for a time when your destination will be some way 'ahead' of your departure point – otherwise, you might have to add the best part of an orbit round the Sun to your journey.

Planning your flight

Although each of the chapters on specific destinations opens with some detailed travel tips, it's a good idea to understand the basics of interplanetary navigation too. There are some cowboy operators out there, just waiting to sell you an 'off-peak' bargain that will turn into the holiday from hell, and you'll save yourself a lot of time and trouble if you can see through any dodgy offers straight off.

As we said before, every destination in the Solar System, the Sun aside, is a moving target. Because every world goes around the Sun at a different speed, the distances between them can change enormously depending on where they sit on their respective orbits. The best example is Venus, which can get to within 42 million km (26 million miles) of Earth at its closest approach, but is up to 257 million km (160 million miles) away when we're on opposite sides of the Sun. The fairly elliptical orbits of some planets add another complication – for example Mars's 'close approaches', when Earth and Mars line up on the same side of the Sun, can vary between 56 million km (35 million miles) and 98 million km (62 million miles)

Earth moves round its orbit at about 30 km/s (19 miles per second). It's faster-moving than anything further out, so once you've broken out of Earth orbit, you'll already have the edge on speed over the outer planets. But of course

LOST IN SPACE

Good flight planning isn't just an exercise in saving fuel and money – sometimes it can be a matter of life and death. Get the timing wrong on a course-changing engine burn and you could find yourself stuck on a trajectory that sends you heading into the Sun or drifting into the outer Solar System. If your fuel reserves are low, you might have no way to get back.

you'll want to get to your destination as quickly as you can, and any extra speed you can pick up will help you to do that. Details of your various propulsion options are available in the 'Traveller's Reference' section (pp.204–220), but there are a number of general points to bear in mind.

First off, it's generally more efficient to accelerate for a longer time with less thrust, rather than piling on all your speed in one burst from an old-fashioned chemical rocket engine (there are health benefits in this too, as we'll see).

Second, don't forget that you can get a significant speed boost from the gravity of other planets you pass on the way to your destination (see 'Gravitational slingshots', opposite). Some seemingly bizarre detours can actually shorten your journey time by months or even years.

INTO ORBIT

The basic principle of orbits is easy once you 'get' it. It's just a matter of balancing your spacecraft's movement against the pull of gravity towards a planet. In any elliptical or circular path around an object, you can split your movement into an element that points along your orbit, and one that points away from the planet. A stable orbit is one where the outward element of your motion is perfectly balanced against the inward tug of gravity. The two cancel each other out, leaving you with just the movement along the orbit. The closer you are to an object, the stronger its gravitational pull, so the faster you will have to move to keep in orbit. Once you're in a stable orbit, though, you should be able to hang around indefinitely without using your engines. To land, just turn around and fire your engines to decelerate and drop towards the planet.

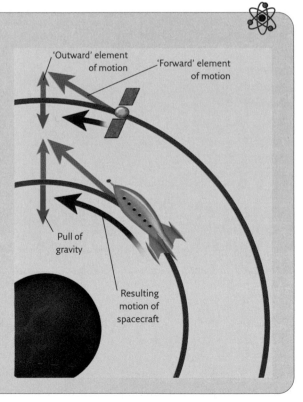

'Outward' element of motion

'Forward' element of motion

Pull of gravity

Resulting motion of spacecraft

GRAVITATIONAL SLINGSHOTS

This popular navigation technique allows you to get a major speed boost on your journeys between planets, without burning any extra fuel. The alignment of the planets is all-important, though – you need to approach your slingshot planet from 'in front' of it, with your eventual target further 'ahead' on its orbital track (see illustration). As you fall into the first planet's gravitational field, you'll pick up speed, as you'd expect. But as you swing behind it and reverse your general direction of travel, you'll find that you don't decelerate as much as expected on your retreat from the planet.

It sounds like a con, but it really works – what's actually happening is that you're stealing some of the planet's own momentum along its orbit. But since the planet is so big (and momentum equals mass times speed), it only slows down by a tiny amount, while your lightweight spacecraft gets a huge boost in speed.

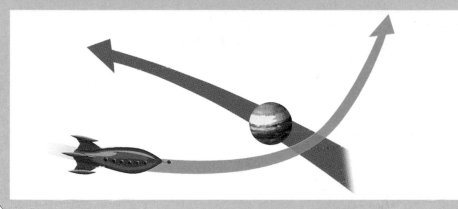

Third, remember that any speed you build up on your journey, you're going to have to burn off in order to match the speed of your destination world and slip into its orbit (unless you're content with going all that way for a brief, high-speed fly-by). There are various ways of doing this (see 'Traveller's Reference', pp.204–217), but try to find a way that keeps the extra fuel required to a minimum. You'll already find yourself carrying a lot of spare fuel all the way to your destination, because...

Fourthly, remember you've also got to get home again! If you burn away all your speed when you arrive at your target, you're going to have to pile it all back on again to get home in a reasonable time. Plan to carry a substantial amount of extra fuel just to break orbit and speed you on the way to Earth.

WHEN SLINGSHOTS GO BAD

Planning a slingshot is a highly specialised skill. Fall into a planet's gravity at the wrong angle and you could find yourself trapped in orbit. Change your escape course at the wrong moment and you could be flung away at a wild angle. If in doubt, get an expert to do the calculations for you!

Using your spacesuit

Apart from your spacecraft itself, the most important piece of equipment you'll carry with you is a spacesuit. In many ways, it's actually a person-sized spacecraft anyway, offering protection and life support (if not exactly comfort) to its occupier.

Most of us can't afford bespoke tailoring, but you can rent or buy 'off the peg' at most reputable suppliers. All modern spacesuits are modular, so you should be able to mix and match bits to fit your shape.

Putting on a spacesuit is a bit like being dressed by your mother: there are an awful lot of layers involved. First comes insulating underwear with several metres of piping running through it – when properly hooked up, your backpack will drive water through these pipes, carrying heat away from areas exposed to sunlight, and releasing it safely elsewhere. It's a bit like wearing a fridge.

On top of that comes a tight-fitting pressure garment, with an airtight membrane surrounded by an elasticated bodysuit. Unless you work out a lot, it won't look flattering, but it's designed to keep your body under the same kind of pressure it experiences from Earth's atmosphere. That way, your bodily fluids won't get any ideas about, say, boiling away into the vacuum of space (see 'Death by space').

Finally, there will be looser-fitting layers of thermal insulation and protective woven material to protect you from any pesky micrometeorites that might try to penetrate your suit. These layers used to be woven from the same stuff they used for bulletproof jackets. Today, they're more likely to be some fancy new 'smart material', designed to slow down any impact, and then revert to its original shape.

Once you've got that lot on, there's still the helmet and backpack to struggle into. The backpack hoses link up to the colour-coded sockets on your suit. Switch it on and it should look after itself, providing you with air- and water-based cooling, oxygen to breathe, drinking water and communications. If you're travelling on your own, then you'll need to

Early spacesuits had problems holding themselves in. When Alexei Leonov made his historic first spacewalk in 1965, he found that his suit expanded so much that he could barely fit back through the spacecraft hatch.

DEATH BY SPACE

The effects of a leaking spacesuit have often been exaggerated – your body won't explode, for instance. In some ways it doesn't matter, but it's nice to know how you're going to go! The first thing you'll experience is the air being sucked from your lungs into the vacuum, so your most likely cause of death will be asphyxia. If you can hold your breath, you'll find that your bodily fluids rise to the surface of any exposed areas of skin. Once there, they'll either boil away in the sunlight or freeze in shadows. It's a nasty way to go, so unless you have an imminent hope of rescue, we'd advise just breathing out and going with the flow...

take a full spacesuit servicing course, but if you're with a guided party, you can just take the short familiarisation course and then do exactly what the nice man tells you. Don't get any clever ideas, as they're likely to get you killed!

Finally, it's on with the helmet, which should clamp into place quite easily. If you can't see anything from the inside, then you've probably got the Sun visor down!

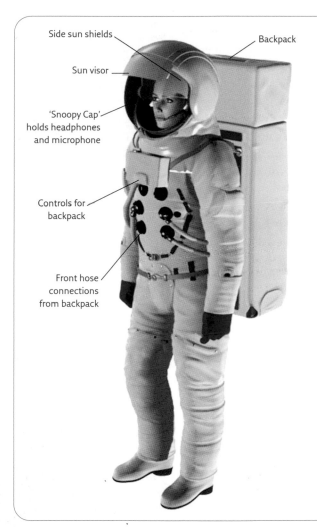

Side sun shields

Sun visor

'Snoopy Cap' holds headphones and microphone

Controls for backpack

Front hose connections from backpack

Backpack

READY FOR ACTION

The right design for your spacesuit will depend a lot on where you're going. Martian suits can be reasonably comfortable and lightweight, while venturing out on Venus is likely to involve dressing up like some kind of cartoon Japanese battle robot. The majority of worlds in this book, however, are airless or near-airless balls of rock. For these, you'll probably want the kind of suit shown here (it's also good for spacewalking and orbital repair work).

Donning a spacesuit is a complicated business, with plenty of room for potentially dangerous errors. If you can, it's a good idea to get a friend or fellow traveller to help.

Health advice

Long-duration spaceflight is a risky business, and despite all the advances in shielding and space medication, there are some factors you really need to keep in mind throughout your journey.

For one thing, you're going to be outside Earth's atmosphere for a long time. Insubstantial though our planet's air may seem, it's amazing how a couple of hundred kilometres of it can mount up, creating a far more effective shield than anything you could carry with you. Earth's magnetic field also helps to create a sort of protective cocoon.

However, once you're outside the atmosphere and beyond the magnetic field, you'll be prey to a wide variety of particles and high-energy radiations that zip around inter-planetary space. A lot of them will pass straight through you, doing no harm at all, but occasionally one will make a direct hit on a cell, either damaging, killing, or mutating it. Shields and medication can only do so much, and by keeping the bulk of your spacecraft between you and the Sun, you can reduce the danger to a minimum.

The other big risks come from prolonged weightlessness. While the initial side effects are relatively easy to get over (see 'Up, up and away!', pp.20–27), the long-term problems are harder to deal with. Muscles rapidly get weaker unless you exercise – on Earth, even the most dedicated couch potato is constantly doing exercise just to hold themselves up against gravity. After a few weeks in space, even bones get weaker – the lack of gravity scrambles the signals that tell them to keep renewing their calcium frameworks.

Away from Earth, your blood no longer sinks towards your feet and ends up much more evenly distributed around your body. Unfortunately that cons your heart into thinking you've got enough oxygen in your blood already, so your body will reduce the number of new oxygen-carrying red cells it makes. It's not a problem in space, but land on any world with a half-decent amount of gravity and you'll risk

Space pioneer Valery Polyakov spent more than 437 days on the Soviet space station Mir in the 1990s, suffering in the cause of space medicine.

CABIN FEVER

Before committing to a long-duration spaceflight, watch some reality TV! If you think the atmosphere gets bad when a bunch of strangers have been cooped up together for a couple of weeks, think how much worse it could get with even less privacy, no escape, and the ever-present threat of cold, vacuumy death. Even astronauts with years of training and experience have been known to go a bit loopy, so pick your travelling companions carefully and use counselling software – it might feel weird confessing your inmost thoughts to a computer, but it's still better than flipping out and trying to evict someone through the airlock.

fainting as your thinned-out blood makes a bee-line back to your boots.

The best solutions are regular exercise and, if possible, artificial gravity (see p.212). Elastic shock cords attached to your belt can give you some 'gravity' to work against when you're on the treadmill or cross-trainer. Russian cosmonauts used to wear spring-loaded overalls that would fold up unless you constantly kept your muscles tense and there are still modern equivalents around. They're not to everyone's taste, though: one lapse in concentration, and you can find yourself in any number of uncomfortable and embarrassing poses.

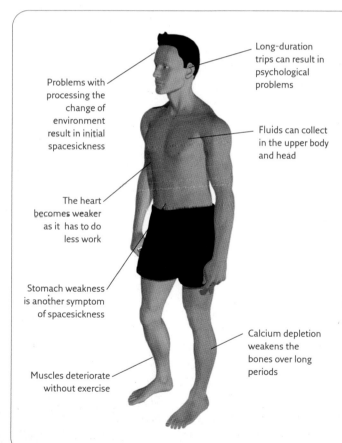

Problems with processing the change of environment result in initial spacesickness

The heart becomes weaker as it has to do less work

Stomach weakness is another symptom of spacesickness

Muscles deteriorate without exercise

Long-duration trips can result in psychological problems

Fluids can collect in the upper body and head

Calcium depletion weakens the bones over long periods

Space Ailments

Long-duration spaceflight is not for hypochondriacs – you'll soon discover that you have more than enough real problems to worry about, without making up new ones. It's important to be in good shape before you go into space in the first place – there are plenty of problems that may seem minor on Earth, but can soon worsen with a dose of zero gravity.

If you've got room for a treadmill on board, it'll help you keep in shape, though you'll have to strap yourself down before exercising.

From the start, let's be honest: getting into space is the difficult bit. The first couple of hundred kilometres on the long road to the Solar System are like trying to push an elephant up the side of the Empire State Building. Actually, that's just the physics. As an experience, it's more like sitting inside a pneumatic drill.

Into the blue

Only a very few people will say that launch is their favourite part of spaceflight, and they're either masochists, or trying to sound tough. If your idea of fun is having someone pinch both your cheeks and make you go 'blubble blubble blubble' before punching you in the stomach very hard, then maybe it'll appeal to you. Otherwise, it's a necessary evil that we all have to go through to get to the fun parts of spaceflight.

Most launches still take place at a handful of spaceports whose origins predate the Space Age. Baikonur and Cape Canaveral/Kennedy (it seems to change its name on an annual basis) both lie as close to the equator as possible in

G FORCES

We'll probably never do away with the large 'g forces' involved in a space launch, and most people are mildly horrified when they see the involuntary facelift you'll be subjected to in the first few minutes of your flight. But there's nothing much to worry about – most people get through it with little more than a few aches and pains. However, we recommend removing contact lenses and false teeth before launch!

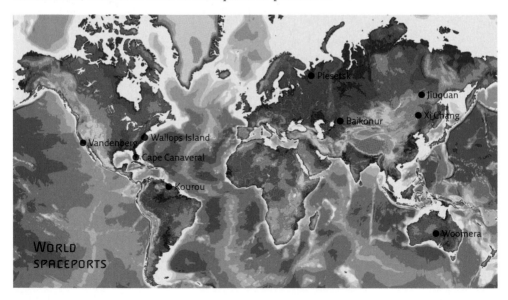

WORLD SPACEPORTS

Plesetsk
Jiuquan
Baikonur
Xi Chang
Vandenberg
Wallops Island
Cape Canaveral
Kourou
Woomera

their respective countries, which means launches can take maximum advantage of the speed boost offered by Earth's rotation. Kourou in French Guiana is much closer to the equator – unhampered by the geography of the Cold War, it was developed in the early days of the European space program. More recent commercial spaceports have also brought their benefits and drawbacks to equatorial nations: with fuel savings of up to 20% on offer, it's hard to resist.

Of course, the days of launching yourself away from Earth in a single rocket burn are long gone. For all its sound and fury, the initial launch is just the messy business of getting you that first couple of hundred kilometres, out of Earth's atmosphere and into Low Earth Orbit. From here, you can spend some time acclimatising – perhaps stopping off at one of the zero-g hotels so you can get to grips with being a spaceperson before hooking up with your ride at one of the various orbital marinas high above the equator.

GOING BEFORE YOU GO

Launch is the riskiest part of any spaceflight, so you'll probably be wearing a full spacesuit, with helmet close at hand, from the moment you board the shuttle bus to the time you achieve orbit. Bear in mind that it's going to be a nervous couple of hours or more, so make sure you stop off at the rest room before you're sealed up with little hope of relief!

DOING YOUR HOMEWORK

Anyone who's venturing into space for the first time should at least try the effects of zero gravity first. Most people adapt quite quickly, but some never get used to it – and it would be a shame to spend all that money on getting to orbit, only to discover that it's not for you. Of course, we recommend that you take a basic space familiarisation course first anyway, but the least you should do is book up for a supervised parabolic flight. Known fondly as 'vomit comets', these specialised aircraft have been plying their nauseous trade since the early days of the Space Age. They're simply converted passenger jets, with the seating area ripped

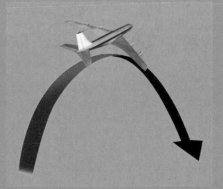

out and lined with padding so it resembles a toddler's playpen (or a padded cell). The plane rises to high altitude, then plunges into a near-vertical dive. As it tilts downwards and the floor drops away from under your feet, you and your fellow inmates will briefly be falling at the same rate as the plane itself, and will experience a precious few moments of weightlessness. Then the plane climbs back up, and you get to do it all again – up to 40 times in the course of one flight. If your stomach's not ready for the rigours of space travel by the end of that, it never will be!

In orbit

It's true what they say: you never forget your first moments in orbit. As the engines cut out and the explosive bolts that release your upper stage send a 'clunk' reverberating through the spacecraft, you'll be freed from the tyranny of gravity for the first time in your life.

Reactions to the sudden disappearance of your entire body weight vary from person to person. Some are elated, most a little nauseous, though today's spacesickness medications should calm all but the most turbulent stomachs. Try a few experiments from the safety of your launch couch before unbuckling yourself and floating free. The biggest challenges of weightlessness are often psychological – your brain is so used to the presence of gravity that it often can't believe what it's seeing. Your stomach is used to a little help from old 'g' too, but this isn't the big issue that the early space pioneers feared it might be: peristalsis, the system of muscle contractions that act as traffic police on the one-way system of your digestive tract, still works perfectly well in orbit.

Weightlessness makes navigating around a cramped spacecraft cabin very different. In theory, now you're no longer stuck to the floor you'll have more room to move. In practice, though, you'll have to worry about a whole load of extra obstacles that were previously kept safely overhead by gravity.

Spacewalking over the night side of Earth is an unforgettable experience for first-time astronauts.

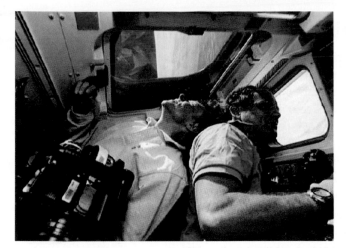

It takes some time to adjust to the fact that there's no 'up' in orbit. In fact, most spacecraft circle Earth with their heavily shielded undersides facing outwards, so you'll usually find Earth itself 'overhead'.

Once your stomach has settled down, don't be ashamed to be a tourist. Weightlessness is fun, so go ahead and have a laugh. Get your travelling companions to take a few snap-shots of you balancing on one fingertip, going head over heels in mid-cabin, or recreating that famous space-docking scene from *Diamonds are Forever* with a slowly spinning banana.

Earth orbit is the ideal place to try your first spacewalk. Floating 200 km (120 miles) up with nothing between you and sudden death but a few layers of material isn't for everyone, but if you can get used to the idea, it opens up a lot more options for the intrepid tourist. Trying it out close to home also means there will (usually!) be someone around to rescue you if it all goes pear-shaped. However, that's no reason to take dumb risks. If weightlessness inside a space-craft is tricky, then manoeuvring outside is ten times tougher. Most insurers will only cover you for tethered spacewalks beyond Earth orbit – there are other options, ranging from miniature manoeuvring guns (you fire one way and move in the opposite direction) to full-blown rocket packs that will allow you to venture much further from your ship. The golden rule, though, is always to keep an eye on your fuel levels, and don't think you can risk running into that emergency reserve!

FLUIDS: A WORD OF WARNING

Fluids – bodily and otherwise – spell trouble in weightless conditions, so make sure you keep them under control at all times. Without gravity to keep it inside containers, anything water-based tends to pull together through surface tension (a force created by the water molecules tugging at each other). To get as close to each other as they can, the molecules form themselves into a spherical shape. This is all very well, but floating blobs of liquid have a nasty habit of getting where they're not wanted – particularly into vulnerable bits of shipboard circuitry. If you're going to lark about with your morning cup of coffee, try not to do it close to the electronics, or shortly before a vital engine burn.

Earth from orbit

Travellers often spend their first few hours in space furiously snapping photos from the cabin window.

Our planet is a dazzling sight from space, and it's worth taking some time while in orbit to appreciate it. Of all the worlds in the Solar System, ours is the only one with extensive oceans on its surface, and of course the only one we know of that is a haven for life. From orbit, Earth is a colourful mix of blue, white, brown and green – hues you'll have to go a long way to see elsewhere in the Solar System. Check the Earthside weather forecasts and see if you can spot any good-looking storms at mid-latitudes. Alternatively, if your orbit has a higher inclination that brings you closer to the poles, keep an eye out for aurorae. These dancing curtains of light ringing the magnetic poles are manifestations of our planet's magnetic field, created where charged solar-wind particles blowing out from the Sun pour down into Earth's atmosphere.

Earth is continually sweeping up other material too, and some of this can present a real risk to the intrepid space tourist. Interplanetary space is full of junk, ranging from microscopic grains of dust to rocks the size of Manhattan. While Project Spaceguard has done its best to chart the ones that could make it through the atmosphere and gouge a city-sized hole in Earth's surface, there's a lot they can't cover, and a head-on collision with even a small chunk of

Even simple cloud formations can be astoundingly beautiful when seen from above. These plateau-like mountainous formations are in fact enormous thunderstorms.

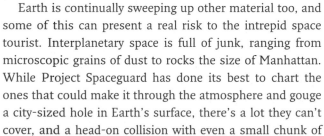

debris could mean an abrupt end to your travel plans, and possibly your very existence.

To be honest, though, there's nothing you can do about it, so it's probably not worth worrying. Should you happen to be in orbit during a meteor shower, our advice is to make your peace with any deities you subscribe to, then sit back and enjoy the unique view.

The sight of a meteor shower from above Earth's atmosphere can be beautiful but unnerving for inexperienced travellers.

SPACE JUNK

For a time back in the early days of the Space Age, it looked as if mankind was going to cut itself off from the wonders of the Universe when it had barely begun to explore them. Early spaceflights were not overly concerned with keeping the environment tidy, under the mistaken impression that space was so big they didn't need to worry about litter. But Low Earth Orbit is actually quite cramped, and by the late twentieth century, astronauts in orbit found they had more than 100,000 pieces of assorted junk for company, ranging from enormous rocket stages to lost tools and jettisoned rubbish. It was around this time that an early Space Shuttle flight was hit by a fleck of paint travelling at several thousand kilometres per hour in the other direction. The shuttle windows were built to take this kind of thing, but the miniature crater left behind was an ominous warning that Low Earth Orbit could become unusable unless we started taking more care.

a giant leap into adventure!

MOON

From the Earth to the Moon

Breathtaking mountains, beautiful seas, but unfortunately no atmosphere... the **Moon** is one of the Solar System's most popular holiday destinations, and there's plenty to see and do. Visit the historic Apollo 11 landing site and have your photo taken with Neil Armstrong's footprint. Dig for the _2001_ monolith in Tycho crater. See the stunning Mare Occidentale, where the Earth hangs forever on the horizon. Play golf in one-sixth gravity!

The Moon has always been the most popular destination for short-haul space tourism. Ever since the first Americans arrived there in the 1960s, it's had a great appeal to space-farers. It's literally on our cosmic doorstep, and surprisingly underdeveloped, although it is (pardon the pun) a little lacking in atmosphere.

Getting there

It's still quite a way from Low Earth Orbit, though, so you'll need a powerful upper rocket stage to set you on course. In the old days, the journey took three days, but of course it depends on the rating of your launch vehicle and the weight of your spacecraft. Keep the weight down if possible, and bear in mind that you can always stop over at one of the established moonbases to pick up extra supplies (the cost of ferrying material to the Moon is exorbitant, so prices are high, but it still works out cheaper than packing everything but the kitchen sink yourself). However, the Moon is an unforgiving environment, so you'll need a spacesuit with all the trimmings – it's as unprotected up there as it is in open space, and you'll have the additional hazard of gravity and sharp rocks to contend with.

A multi-stage spacecraft will be economical here, as it is for most Solar System destinations – you can park the command module in orbit, and use a descent module to come and go from the surface. That way, you'll avoid taking unnecessary weight all the way down to the lunar surface, just to bring it back up again. The Apollo astronauts used the biggest rockets ever built to get from the Earth to the Moon, but by following this mission design, they were able to make the return voyage with a couple of low-powered rocket engines and a comparative smidgeon of fuel – and there's no reason why you shouldn't do the same.

Another thing to bear in mind is that you'll be crossing the Van Allen radiation belts on the way. These are doughnut-

LUNAR DATA

Good points:
Close to Earth.
Low gravity fun!

Bad points:
Lacks atmosphere.
Craters can get monotonous

Day length:
27.3 Earth days

Year length:
1 Earth year

Gravity:
0.17g

Surface temperature:
-150 to 120 °C
-240 to 240 °F

Communications time:
About 1.3 seconds

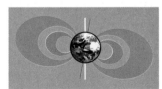

The Van Allen Belts loop around the Earth, like two deadly ring doughnuts filled with fast-moving, highly penetrating particles.

ILLUSIONS AND ERUPTIONS

The Moon has a particular reputation for playing tricks on observers who've spent too long with an eye to the telescope. Even the great eighteenth-century observer William Herschel wasn't immune. In 1787, he reported the discovery of what he thought was an erupting volcano on the Moon, but it turned out to be the bright crater Aristarchus glowing in particularly bright Earthlight. However, a lot of astronomers agree the Moon sometimes shows genuine activity. Transient Lunar Phenomena are occasional orange glows that come and go from the lunar surface, and seem to concentrate around relatively young craters. They're frustrating little blighters, since their appearances are so brief and unpredictable that catching them on film is the astronomical equivalent of whacking a fairground rat, but they do seem to be a genuine phenomenon. Selenologists (Moon experts) are pretty certain that they're caused by gas pockets escaping from below unstable ground, perhaps set off by small meteorite impacts or moonquakes.

shaped regions of Earth's magnetic field that trap particles from the solar wind, and send them ricocheting back and forth between the Earth's magnetic poles. You'll have to spend a few hours passing across them, so make sure your spacecraft is well shielded. Even so, be prepared for some unusual side-effects – a solar wind particle passing through your head can cause the nerve cells at the back of your retina to fire, causing your brain to 'see' a bright flash.

When you reach lunar orbit, be sure to spend some time enjoying the view. A leisurely orbit may take several hours, giving you plenty of time to appreciate the varied lunar terrain. It's a common myth among those who haven't been that the Moon is 'just craters', but that's unfair. Even from Earth you can tell that there are two main types of terrain – the bright highlands (which, to be honest, *are* mostly craters), and the darker grey *maria* or 'seas'. These occupy lowland areas, often filling up the ghosts of huge, ancient impact basins.

Keep a look out in particular as you cross the terminator line separating lunar night and day. With no atmosphere to scatter the sunlight, the plunge into darkness comes suddenly, and mountain ranges and crater rims can look spectacular as they catch the light of the setting Sun in an otherwise dark landscape.

KONSTANTIN TSIOLKOVSKII

Spare a thought as you pass over the Tsiolkovskii crater. It commemorates the life of Konstantin Tsiolkovskii, a Russian schoolteacher from the turn of the twentieth century who was first to prove that rockets would work for space travel. He also worked out much of the nitty-gritty physics of spaceflight. Without his work, you might not be able to see the crater for yourself.

The Sun isn't the only source of light on the Moon – the nearside gets the benefit of Earthlight too. Our planet reflects far more light onto the Moon than the Moon bounces onto Earth, simply because of its size, and this light often washes the night-time moonscape in a pale grey-blue.

Of course only half the Moon benefits from the light of the Earth – the other half is perpetually turned away from its parent planet. Cross the line from near to far side, where the Earth sinks below the horizon, and watch the wave of darkness sweep beneath you – night on the far side of the Moon is blacker than black, lit only by stars that appear more spectacular than the darkest night on Earth, even through a heavily shielded visor.

Whatever Pink Floyd might say about it, though, there is not really a 'dark side of the Moon', and the far side receives just as much direct sunlight as the near side. Early rocket engineers went to tremendous lengths to get the first photos of the sunlit far side, but to be honest it's something of a disappointment – there are very few of the seas that are so prominent on the nearside, and the rest is a confusing mass of highland craters, often with unpronounceable Russian names. Two areas that may help you get your bearings are the Mare Moscoviensis, the far side's best effort at a proper lunar sea, and the dark-floored crater Tsiolkovskii.

LUNAR TIDES

The Earth and Moon are both affected by tidal forces. Each world's surface is tugged towards the other's, creating a bulge pointing directly towards the other body. A second 'tidal bulge' forms on the opposite side, counterbalancing the first. As a planet or satellite spins on its axis, tidal forces will tend to slow it down until the bulges no longer move relative to the object causing them. That's why the Moon has now slowed down so one side permanently faces the Earth – a phenomenon called 'synchronous rotation' that we'll see among many of the Solar System's satellites. Tidal forces also produce a lot of heat inside a body as it flexes its shape (see p.124).

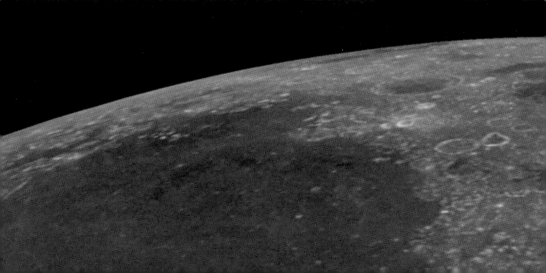

INSIDE THE MOON

The Moon's internal structure is fairly simple. The crust is largely made of volcanic rocks that are similar to the granite found on Earth, and is thicker on the far side than on the near side – one reason why later volcanic eruptions, such as those which formed the *maria*, were more common on the near side . Beneath this is a deep mantle of solid rock. The moon weighs too much to be made of rock alone, however, so it seems there's a small core of nickel and iron at the centre. Unlike Earth's core, however, the Moon's has long since cooled and solidified.

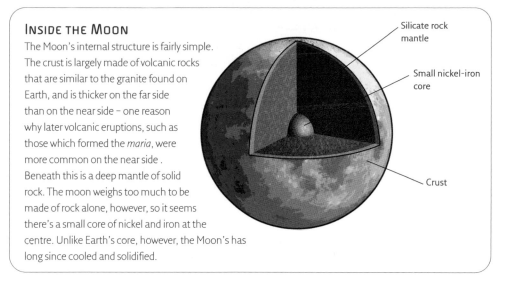

Silicate rock mantle

Small nickel-iron core

Crust

A couple of lunar orbits is probably enough to satisfy all but the most hardened luna-tics; after that the landscape starts to get a bit monotonous. Think how it must have been for those poor Apollo Command Module pilots, stuck in orbit for days while their travelling companions were living it up on the lunar surface! After a few hours, you'll be itching to get down among the rocks, and where better to start than the place where it all began?

You can enjoy this spectacular view from low orbit for yourself as you speed over the Mare Crisium on your way towards the lunar farside.

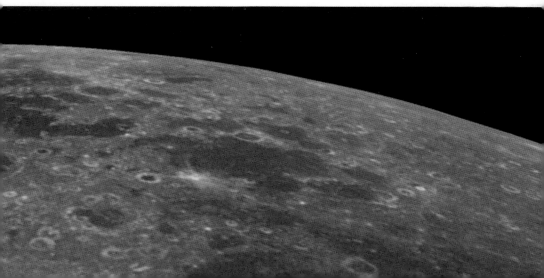

Tranquillity Base

Tourism in the Solar System isn't like tourism on Earth, so don't let snobbery get the better of you – just because everyone else goes somewhere, it doesn't mean that you shouldn't too. The Sea of Tranquillity is a case in point: if you visit the Moon, go and see it, period. Not only is it a historical landmark, the site of the first Moon landing, but the views are spectacular. It's smack in the centre of the Moon as we see it from Earth, so our home planet hangs permanently overhead as it shifts through the endless cycle of phases in tune with the lunar orbit of just under four weeks. The landscape is relatively featureless, a rolling plain of titanium-rich volcanic rock, spattered with small craters and the rocks ejected as they formed. Bear in mind that the NASA experts planning the trip had little idea what to expect (they'd only just convinced themselves that a spacecraft landing on the Moon wouldn't sink into the dust, never to be seen again!) so they can be forgiven for hedging their bets by not aiming for a more 'exciting' (read 'dangerous') area of the lunar landscape.

Michael Collins got the short straw on Apollo 11 – as Command Module Pilot, he had to stay in orbit while Neil Armstrong and Buzz Aldrin had fun on the surface.

But if you're coming here, let's be honest, it's not for the view. The landing site where Neil Armstrong and Buzz Aldrin first set foot on the moon is still in pristine condition – and likely to remain that way for millions of years unless a chance meteorite gets lucky. Unfortunately, thanks to concerns about souvenir hunters stealing memorabilia and putting their size ten moonboots all over the place, all six Apollo sites have to be protected with 20-metre tall fencing (extra high to deter low-gravity catburglars). However, there are good viewing platforms, and some of those epochal footprints are outside the fence and protected by plexiglass, so you can still have your photo taken standing in Armstrong or Aldrin's footsteps.

The most spectacular artefact preserved here is the spider-like lower stage of the lunar lander. It still bears the scorch marks where the upper section blasted free to take the astronauts back home at the end of their brief stay. It's salutary to think that Armstrong and Aldrin put up with six days in the

Fortunately the footprints are now protected, but you can't help worrying when you're walking on top of them.

cramped confines of the Apollo spacecraft for just 21 hours on the surface of the Moon, and even more sobering to remember Michael Collins, the Command Module pilot, who went all that way only to babysit the spacecraft in lunar orbit.

Nearby sits a package of scientific instruments set up by the astronauts. Most were turned off during one of NASA's regular cost-cutting exercises in 1977, but Earth-based scientists still occasionally bounce lasers off the reflective panels of the Lunar Laser Ranging Experiment to re-measure the gradually increasing distance to the Moon. A word to the cost-conscious here: go to the Moon as soon as you can, because every 26,000 years you wait, you'll have to go one kilometre (0.6 miles) further! The Moon's slow and steady outward spiral is a result of the same tidal effects that have created its synchronous rotation, and that have even slowed the Earth's spin by a couple of hours since the time of the dinosaurs.

One last highlight is the plaque that the astronauts placed to commemorate their visit. It's worth reciting the text in full:

Here men from the planet Earth
First set foot upon the Moon
July 1969, A.D.
We came in peace for all mankind.

THE FLAG

A lot of effort went into selecting the Apollo 11 flag that Armstrong and Aldrin took to the Moon. NASA bought in flags from a wide variety of manufacturers, but they didn't want anyone making commercial capital out of their product going to the Moon (a far cry from this era of sponsored rocket launches!). The solution? The flags were shorn of identifying labels, and the choice was given to a secretary left alone in an office.

Flying in towards the Tranquillity Base Visitor Center takes you over the historic landing site itself.

The Apollo tour

Apollo 11 gets all the glory, but don't forget that those early US astronauts got around a bit. Trips to take in each of the Apollo landing sites are an ideal way to get a flavour for the variety of terrain found across the Moon. Particular highlights include:

● Apollo 12, Oceanus Procellarum. The second manned landing on the Moon touched down in the Ocean of Storms, a lunar sea blanketed by the bright 'ejecta' material from several craters (the major crater Aristarchus lies just to the east). The landing site is a gentle stroll from the earlier Surveyor 3 robot probe. The Apollo astronauts came here to see how the earlier visitor had fared after a couple of years on the lunar surface: not badly, as it turned out, and it's still in good nick even today.

● Apollo 14, Fra Mauro. Originally intended as the landing site for Apollo 13 (the lunar mission that almost ended in disaster and subsequently became the most famous of them all), Fra Mauro is a hilly area blanketed in material that was thrown out by the formation of the nearby Mare Imbrium. This is a moonquake zone, an area where the surface is relatively unstable and occasionally slips under the influence of Earth's gravity, especially around monthly perigee (the closest approach to Earth). It's hardly the Los Angeles Canyon, though – the quakes are so small that you'll be lucky to even feel them.

Harrison Schmitt loads up the Lunar Rover during the Apollo 17 mission. The rovers are still there, but they've been disabled, so don't try joyriding!

DUST, ANYONE?

One thing that everyone says about the Moon is that it's messy – the dust gets into every nook and cranny (and we mean *every* nook and cranny). That's partly because it's so fine – billions of years in a vacuum, being pulverised by micrometeorites, will do that to a landscape. Back in the mid-twentieth century, astronomers had no idea how deep the dust went. They worried that the lunar landscape might be as stable as a giant sandcastle, nothing but dust for several miles down. Fortunately, the first soft-landing lunar probes showed this wasn't the case: there are plenty of larger rocks, and they do a good job of holding the lunar soil or 'regolith' together.

Still, there's an awful lot of dust, and it's very clingy (partly due to static electricity that builds up as the soil is pummeled by charged particles in the solar wind), so don't get hung up on keeping your spacesuit pristine and white. For all the hygiene precautions the Apollo astronauts took, they still ended up being able to smell and taste the Moon rock once they were back inside the Lunar Module.

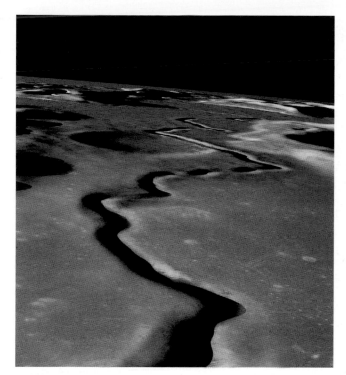

Here, we're flying above Hadley Rille, the gigantic collapsed lava tube visited by the Apollo 15 crew. Spot the crater!

Golf tips

Keen golfers may want to honour Alan Shepard, Commander of Apollo 14 and the first person to tee off on the Moon. He took three tries to get a decent stroke, and that had to be marked down as a lost ball. Unless you want a very long and fruitless search, we recommend you stick to putters and fit your golf balls with a tracking device. There are plenty of natural bunkers to avoid, but at least you won't have to worry about water hazards.

Apollo targets

Apollo 15
Hadley Rille

Apollo 17
Taurus-Littrow

Apollo 11
Mare Tranquillitatis

Apollo 12
Oceanus
Procellarum

Apollo 14
Fra Mauro

Apollo 16
Descartes

● Apollo 15, Hadley Rille. Probably the most interesting site after Tranquillity Base, a visit to Apollo 15 offers the first of the famous Lunar Roving Vehicles (the earliest car to drive on the Moon) as well as Hadley Rille. This is one of the most prominent volcanic features on the Moon, a collapsed lava tunnel more than a kilometre wide.

● Apollo 16, Descartes. The first landing site in the lunar highlands, Descartes is another place you can see a Lunar Rover slowly gathering dust. This is a hilly region, formed when ejecta from nearby major impacts slammed into the original surface and the pulverised rocks melted together.

● Apollo 17, Taurus-Littrow. This valley is actually a plain of *mare*-type lava with highland hills sticking up through it. In places, you can see where the molten lava lapped up against the sides of the hills. You can also see where astronaut Gene Cernan took the last step on the Moon for almost half a century, back in December 1972.

Tycho

You can't go to the Moon without visiting a really good crater, and Tycho is one of the best. At 85 km (52 miles) wide, it has a perfect bowl shape and mountainous central peak. But for all that, it's probably more famous through fiction than fact (see 'Monolith hunting in Tycho', opposite).

Sci-fi aside, Tycho is an impressive place. While on Earth water, wind, and life all act to smooth out the landscape and wear down features such as craters over tens of thousands of years, the only erosion on the Moon comes from stress as the temperature changes, and the fall of other meteorites. As a result, Tycho has remained in pristine condition for roughly 100 million years. It's still surrounded by bright 'rays' of material that squirted out as the impact crater formed, and travelled up to 1,500 km (940 miles) in the weak gravity before falling back onto the lunar landscape. Closer in, you can see the smaller secondary craters where large chunks of ejected material fell back closer to the original impact site.

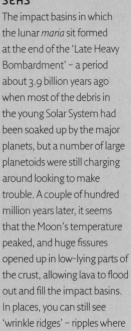

IMPACTS AND SEAS

The impact basins in which the lunar *maria* sit formed at the end of the 'Late Heavy Bombardment' – a period about 3.9 billion years ago when most of the debris in the young Solar System had been soaked up by the major planets, but a number of large planetoids were still charging around looking to make trouble. A couple of hundred million years later, it seems that the Moon's temperature peaked, and huge fissures opened up in low-lying parts of the crust, allowing lava to flood out and fill the impact basins. In places, you can still see 'wrinkle ridges' – ripples where the lava formed custard-like skin before setting properly.

An orbital view of Tycho.

CRATER MAKING

A surprising number of lunar tourists will cheerfully admit to not knowing the first thing about impact craters, and that's a shame. Okay, they're still spectacular even if you don't know how they work, but you get a lot more out of places like Tycho when you understand a little bit more about them. Don't forget that until the start of the Space Age, a lot of astronomers thought the craters might all be extinct volcanoes!

The process behind cratering is pretty simple when you get down to it. An incoming asteroid, comet, or whatever (it's often easiest to use the catch-all term 'bolide') hits a planet or moon at a speed of several kilometres per second, perhaps 10,000–20,000 km/h (6,000–12,000 mph) . As it makes contact with the larger body, a shockwave blasts out from the impact area, pulverising the rocks around it, crushing them together and heating them up. Directly behind the shockwave follows a region of decompression, so the compressed rocks are suddenly able to expand, and material from directly beneath the impact shoots out in high-speed jets in all directions, creating an expanding, bowl-shaped hole in the surface. As the shockwave continues through the crust, it weakens, and so does the decompression wave, so that material from lower depths is thrown out with less force, and falls back to the ground closer to the crater.

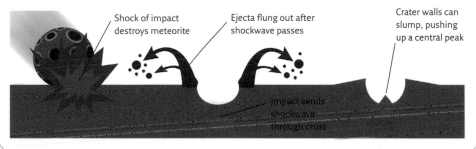

Shock of impact destroys meteorite

Ejecta flung out after shockwave passes

Crater walls can slump, pushing up a central peak

Impact sends shockwave through crust

MONOLITH HUNTING IN TYCHO

Tycho is where, according to the visionary twentieth-century science fiction movie *2001: A Space Odyssey*, aliens buried a mysterious device known simply as the Monolith. This rectangular black slab was in reality a supercomputer, designed to watch over the development of life on Earth and send a signal back to its makers when intelligent life discovered it on the Moon. Although entirely fictional (we think), the idea behind the Monolith is a smart one – a thousand such probes sent to watch over habitable planets would only require a fraction of the resources needed to broadcast a powerful radio signal to say "hello".

A trip to the far side

The back side of the Moon tends to get much less tourist traffic than the near side, for a combination of reasons. Firstly, there's a distinct lack of 'landmark' geography – and where's the fun of going somewhere that you need three hours to find on a map when you tell your friends about the trip? Secondly, the absence of the reassuringly permanent Earth in the sky can be a bit off-putting for the less adventurous traveller. Lastly, there's the lack of communication with home. You might be only a light-second from the folks, but when there's 3,400 km (2,100 miles) of solid rock in the way, you might as well be on the far side of the galaxy.

Earthrise over the Mare Occidentale is an unforgettable sight.

As a result, the lunar far side is mostly left to selenology PhD students and astronomers: without Earthlight to worry about, the night skies here can be more spectacular than anything on Earth, and the gravity allows the construction of telescopes far bigger than anything that they can put easily into orbit. There's also a complete lack of radio noise, except when the emergency comsat passes overhead once every few hours.

THE FAR SIDE OF THE MOON

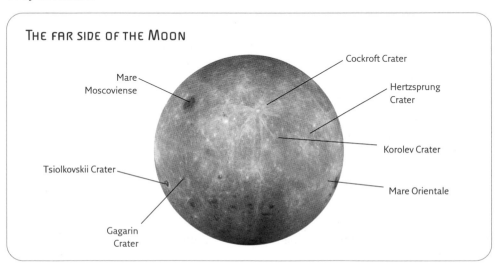

Cockroft Crater

Mare Moscoviense

Hertzsprung Crater

Korolev Crater

Tsiolkovskii Crater

Mare Orientale

Gagarin Crater

But what is there to see? Highlands, mostly: lots and lots of craters and mountain chains, most with tongue-twisting names. The major attraction is probably the aforementioned Tsiolkovskii crater. Its appeal comes from the fact that it's basically a fun-sized *mare*, a high-walled crater 198 km (123 miles) across, its floor covered in dark basaltic rock. For some reason it was here, of all places, that the dark *mare* lava overcame its apparent reluctance to erupt on the far side, and put in an appearance to flood the depths of the crater.

Standing on the rim of Tsiolkovskii, the view is dominated by the dark volcanic plain and bright central peaks.

Tsiolkovskii is also blessed with a pretty spectacular central mountain, which sticks up like an island from the surrounding sea. Stand on the cliff edge at sunrise and the view across Tsiolkovskii captures some of the most beautiful aspects of lunar geography at a stroke.

Snow on the Moon

From Tsiolkovskii, it's a relatively short hop southeast to the far side's other great attraction. Although not as photogenic as some of the smaller craters and basins on the near side, the South Pole-Aitken Basin is a must-see on account of its sheer size. This vast impact basin is the largest in the Solar System, extending all the way from the lunar south pole up to mid-southern latitudes on the far side. With a diameter of roughly 2,500 km (1,550 miles), about the size of western Europe, it even outclasses the giant Caloris Basin on Mercury.

Despite its huge size, the basin is a bit of a disappointment from orbit. Even though the impact, roughly 3.9 billion years ago, gouged out rocks to a depth of 12 km (7.5 miles) below the lunar surface, it didn't strike magma, so the floor of the basin remained resolutely unflooded. This makes it bright and hard to distinguish from the highlands around it.

Most people that come this far south, however, are only here for one thing: the ice. Nestled in deep craters around the south pole are ice deposits left over from the occasional comets that have hit the Moon. Okay, it's generally not dense enough for winter sports, but snowball fights in one-sixth

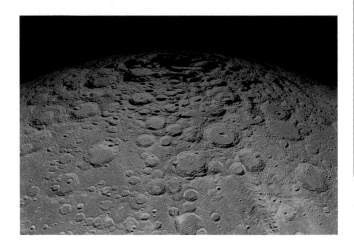

THE BIG SPLASH

The Moon's huge size has always made it a mystery. None of the other inner planets has such a substantial satellite. Its geology is also puzzling – similar, but not the same, as Earth's. The best theory to explain all this is called the 'Big Splash'. Shortly after the Earth formed, a Mars-sized rogue planet (sometimes called Theia after the Greek Moon goddess's mum) came hurtling towards our planet and struck it a glancing blow.

The collision destroyed Theia and blasted off a large chunk of Earth's mantle. A lot of the debris flew off into space, but some settled back to Earth, and a substantial amount went into orbit around our planet. Here it collided and merged together, creating the Moon in perhaps just a few decades.

The enormous Aitken Basin lies in permanent twilight at the lunar south pole.

Ice lurks in a permanently shadowed crater within the Aitken Basin.

Earth gravity can be a lot of fun! More to the point, this ice comes from comets and so it's basically unaltered since the birth of the Solar System itself, 4.6 billion years or so ago. Ponder on that as you wipe it off your helmet visor!

Selenologists first suspected the ice was here when a late-twentieth-century space probe called Lunar Prospector picked up unusually bright radar reflections from some of the craters. They even tried to clinch the case by smashing the probe into one of the craters and watching for a puff of water vapour evaporating as it was thrown up into the sunlight (though without much luck). However, when the second wave of human explorers finally made it to the Moon, the ice was one of their top priorities. Not only is it a handy fuel source, easily split apart to make liquid hydrogen and liquid oxygen, but it's also an archaeological treasure: a plentiful sample of the Solar System's deep-frozen primordial soup, easily accessible on our planetary doorstep, with no need to chase comets around the Solar System.

WHY HERE?

Ice and liquid water survive on Earth because of our planet's thick atmosphere, but on the Moon, they should boil away to nothing in the Sun's heat. So how come ice survives at the poles? It's all down to a curious fluke of the Moon's orbit. While the Earth is tilted at a lazy 23 degrees to the plane of its orbit, meaning that all areas of the planet get a reasonable amount of sunlight for at least part of the year, the Moon's orbit is only tilted at 6.5 degrees from Earth's. The same tidal forces that have slowed down the Moon's rotation also ensure that the Moon stands to attention, bolt upright in its orbit. So at the poles, the Sun only ever rises a few degrees above the horizon, and the walls of deep craters (particularly those already inside the South Pole-Aitken Basin) are often high enough to ensure that the crater floors are hidden in permanent shadow.

Further afield — Cruithne

According to astronomical knowledge going back for centuries, the Earth has only one moon. Just to be sure, we've checked with a number of people who have counted it. But there are several other weird objects out there that are almost satellites of Earth, if you squint at them from a favourable scientific point of view.

Cruithne in particular is often called Earth's 'other' moon. The name's almost as confusing as its bizarre orbit – it's properly pronounced something like 'Crooy-nyuh', although since it comes from the name of an ancient Celtic tribe, and they're no longer around to ask, we'll probably never know for sure.

When astronomers first discovered this strange little world, they thought it might actually be a stray satellite or

HORSESHOE ORBITS

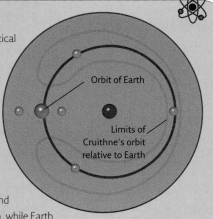

Cruithne's average distance from the Sun is just about identical to Earth's, but its orbit is a bit more elliptical, so it goes from inside our orbit to outside it. (It's also tilted at an angle to our orbit, so there's not much chance of a collision with Earth). For some of the time, Cruithne's year is a little shorter than ours, so it orbits the Sun a little more quickly. From Earth's point of view, it seems to spiral through space ahead of us, getting slightly further ahead with each loop.

Orbit of Earth

Limits of Cruithne's orbit relative to Earth

Eventually, the asteroid starts to catch up with Earth from behind, until it comes as close as 15 million km (9.3 million miles) or so. Around this point, tidal forces between Earth and Cruithne mean that Cruithne loses a little of its momentum, while Earth picks up the same amount (it's a little like a gravitational slingshot in reverse). Because Cruithne is so much smaller than Earth, a little lost momentum makes a lot of difference to its speed – enough to lengthen its year so that it now becomes longer than Earth's.

Cruithne then starts dropping further away from Earth with each orbit, until eventually Earth starts to catch up with it. Once again, when they get within 15 million km of each other, they start to swap momentum, but this time it's the asteroid that picks up speed, and starts to move ahead of Earth again. So Cruithne is doomed by perpetual ping pong to never get close to Earth, which is probably just as well for us.

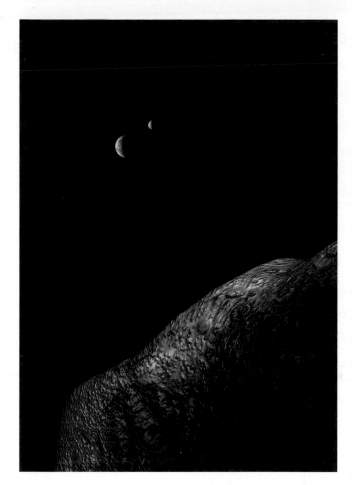

Earth and Moon are just tiny disks in Cruithne's sky, even during a close approach, as in this view from 15 million km (9.3 million miles).

piece of space junk – though the discovery that it's about five kilometres (roughly three miles) long put an end to that idea. It turned out to be a normal Near-Earth Asteroid, but one that's been nudged into an unusual type of orbit (see 'Horseshoe orbits', opposite).

To be honest, Cruithne's nothing to write home about – a typical, crater-scarred ball of rock. However, if you're in the vicinity during a close approach, it's worth dropping by for a spectacular and unusual view of the Earth and Moon in space together.

♥ *VENUS is for lovers*

Tourism at the extreme

Venus is just too much for most holidaymakers. Too hot, too corrosive and too poisonous, its runaway greenhouse effect offers an important lesson for Earth. Are you brave enough to risk descent to the surface in an armoured capsule, and to walk on the surface in a heavily shielded spacesuit? If so, the wonders that await you include the giant volcanoes of Maxwell Montes, the frozen lava rivers of Maat Mons, and the decayed remains of early Soviet spaceprobes. Be careful, though: stay too long and you might never leave!

Although Venus in our nearest planetary neighbour, it's yet to really catch on as a holiday destination. A dangerous reputation gained in the early days of space travel has stuck, not helped by the fate of a few badly organised tour parties that disappeared into its acidic skies and were never heard of again.

But provided you take good safety precautions and have adequate insurance cover, Venus has a lot to offer. It's tantalising to visit a world that's so close to Earth in a lot of its essentials, yet has turned out so differently – and it's also sobering to think that the same fate might one day befall our planet.

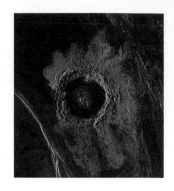

The dense Venusian atmosphere stops craters from flinging their ejecta very far. Instead, it falls down close to the original impact site. This may help to creare lobed craters like Dickinson, which has no ejecta at all on one side.

Getting there

Travelling to Venus can be a joy or a nightmare, depending on the timing. Make sure you check the positions of the planets with your local travel agent before scheduling your trip. Depart at the wrong time, when Venus is on the

MAPPING VENUS

Even today, we'd have no idea of the overall geography on Venus if it weren't for radar maps. Traditionally, radar works by firing 'pings' of radio waves at an object and seeing how long they take to come back. It's a bit like a tennis player working out the distance to a wall by counting how long his ball takes to bounce back, only in this case the ball is travelling at 300,000 km/s (186,000 miles per second) – quite a forehand smash.

A more sophisticated form of radar involves firing a longer 'chirp' of radio waves at the target object. To overstretch the metaphor, it's as if the tennis player hit a half-dozen balls at the wall more or less simultaneously. By working out the time different parts of the chirp take to come back, and the angle at which they return, it's possible to discover its shape in a lot more detail, not to mention other information such as its roughness and composition.

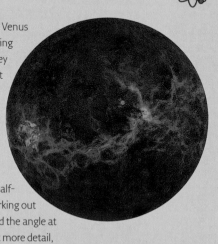

opposite side of its orbit from Earth, and you'll have to complete almost an entire circuit of the Sun just to catch up with the planet. It's time-consuming, expensive and, frankly, boring (there's really not a lot out there in the Earth-Venus gap). Get it right, and you'll shave months off your journey time and benefit from easier communications with home, as the two planets briefly run neck-and-neck around the Sun, barely 42 million km (26 million miles) from each other.

Eventually, though, Venus and Earth inevitably draw apart again, because Venus moves quite a bit faster around its orbit than Earth does. This means you'll have to pick up speed to catch up with it and enter orbit, but this certainly isn't the issue that it becomes for speedy Mercury (see p.60). If your spacecraft can get you out of Earth's gravitational field, you should be able to catch up with Venus, no problem.

During the voyage, test your eyesight by seeing how soon you can pick out the planet's phases. The brilliance of Venus's cloudtops makes the shape of the daylit side easier to spot (the planet has the highest reflectivity of any in the Solar System). Some people can even tell that it's not perfectly circular with the naked eye from Earth. You could also keep an eye out for the diffuse glow that sometimes suffuses the night side (see 'The ashen light' on p.51).

Once you're safely in orbit, you'll probably find yourself sick of the view within hours. The planet's completely covered in cloud and the brilliance of the reflected light washes out any detail. The sulphurous yellow glow it inflicts on everything also gets irritating very quickly. For a less migraine-inducing atmosphere, try any of the various window filters available. Ultraviolet filters are particularly effective at blocking out the yellow and increasing the contrast, so not only will Venus's angry yellow be replaced by a soothing blue (ideal for chilling out after a surface expedition), but you'll also be able to see the huge chevron-shaped cloud systems that encircle the planet, swirling around it once every four days or so.

VENUS DATA

Good points:
Close to Earth.
Rarely visited

Bad points:
Lots of interesting ways to die

Day length:
243 Earth days

Year length:
225 Earth days

Gravity:
0.9 g

Surface temperature:
470 °C
870 °F

Communications time:
140 seconds or more

The planet with atmosphere

Any expedition to the surface of Venus is going to involve getting through the atmosphere first, and that's a risky proposition on its own. Venus is one place where the phrase 'you could cut the atmosphere with a knife' is almost literally true: the air is a hundred times denser than Earth's own. But that's only half the problem, since it's also highly toxic and corrosive. This combination spelled doom for a number of spaceprobes in the early decades of space exploration (around the mid-twentieth century) before their designers figured out just how tough the descent to the surface was. So make sure that all your equipment is safety certified, and don't even think about cutting corners and buying second-hand kit.

Despite the thick atmosphere, you'll need a launch vehicle with retrorockets to slow you down and drop you onto the surface. Parachutes are strictly for the last couple of thousand metres of descent, as they'll start to corrode the instant they're deployed. Back in the early days several probes made it down on parachutes alone, but they didn't have crews to worry about. These days, no insurer would come within a mile of you.

Volcanic peaks loom out of the surface haze during final descent. The main cloud layers hang overhead, permanently blocking out the Sun.

To be honest, this is one of the few points in the book where we're not going to advocate doing-it-yourself. In general, we're great supporters of independent space travel, but the sheer number of delightful ways that Venus can kill you makes it an exception. Take our advice and talk to one of the respected extreme tourism operators.

Lecture over. However you get there, the descent through Venus' atmosphere is unforgettable. Assuming your descent module is fitted with those natty new transparent aluminium windows, you'll be able to see it first-hand. Otherwise, you'll have to make do with the picture from your hull cameras. As you dip towards the cloud layer, the view gets unbearably bright. Don't be ashamed to wear solar filter sunglasses if you have them or they're provided on your flight. The trip through the clouds is the most turbulent part of the descent, and for first-timers it can be as unnerving as a foggy descent into London City Spaceport: you can't shake the feeling that the altimeters have failed and the ground's racing towards you, unexpectedly eager to say 'hello'.

However, Venus's cloud base is surprisingly high. The temperature at ground level simply boils away any vapour that tries to condense below about 45 km (28 miles). As you drop out of the clouds, you'll begin to make out a great bowl of Venusian landscape below, emerging gradually through the yellow haze. Your pilot should know to steer well clear of any volcanoes, but if you're lucky, you may come down within sight of one. You're unlikely to see an eruption (if you do, you'll be the first, and get a place in the history books), but the electrical storms that frequently flare up over volcanic locations are spectacular to say the least.

As you close in on the surface, don't be too alarmed if your vessel's hull starts to creak. The pressure change from the vacuum of space to the surface of Venus is huge – at one hundred times Earth's, Venusian atmospheric pressure is equivalent to a depth of 1 km (3,300 ft) in Earth's oceans. Pressurised spacecraft must be designed to 'give' a little as they adjust from one extreme to the other.

THE ASHEN LIGHT

People have been seeing the ashen light for centuries, and while at first some astronomers dismissed it as an illusion caused by too many nights hunched over a cold telescope, it's now accepted as a real phenomenon. What causes it is a different matter, and today's theories range from storms in the atmosphere to active volcanoes. Earlier explanations (more fun, but now sadly ruled out) included the 'moonshine' from an unseen satellite of Venus, and seasonal fires lit by Venusians to celebrate their harvest festivals. Astronomers used to be so much more inventive!

Stepping out in style

Venusian gravity is roughly 90% of Earth's, which has a number of disadvantages. For one thing, it's so close to normal that visitors often find themselves feeling unusually clumsy as their Earth-honed instincts try to kick in. More of a problem, though, is that Venus doesn't give you much weight advantage to help you cope with the sheer amount of gear you'll have to carry onto the surface. The most over-burdened round-the-world backpacker has it easy compared to someone preparing for a couple of hours on the Sif Mons plains. Add a medieval suit of armour and deep-sea diving boots and you're coming somewhere close

A vista across the volcanic Venusian landscape reveals distant lava flows, cracked rocks, darker 'soil', and the ever-present yellow sky.

to the experience. But is it worth the effort? It depends on how adventurous you feel, really, but you'll have to go a long way out in the Solar System to find a major world where fewer people have stood.

Surface conditions can be treacherous, because so much of the Venusian landscape is made from fairly light volcanic rocks. Almost the entire surface of the planet is covered in volcanic material, and nothing much has happened since they formed to change the consistency of the original lava deposits, so some areas are as slippery as glass, while others are as porous as pumice. Keep the burden of your protective armour in mind before trusting your weight to anything that looks unstable.

VENUS SUIT SECURITY

We can't emphasise enough how much your suit integrity matters on Venus. Elsewhere in the Solar System, a suit failure will kill you in a minute or so, but here it'll all be over in seconds. If in doubt, check, check and check again!

Eistla Regio

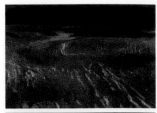

Most people come to Venus for its volcanoes, and they are absolutely spectacular, even when dormant. Venus seems to have invented more ways of getting lava onto its surface than any other planet in the Solar System. There are traditional shield volcanoes of the type you'd find on Earth, 'pancake domes' that form stoppers over ancient volcanic vents, and a wide range of cracks and fissures in the surface. Most spectacular, though, are the volcanoes of the Maxwell Montes region, around the Venusian equator.

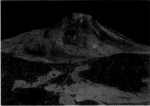

Flying in over the plains around Gula Mons, you'll notice that the ground has a rippled effect, like wallpaper paste left too long in the bucket. In fact, this isn't too far from the truth – this landscape is formed entirely from frozen lava, and the same goes for 90% of Venus. By counting the number of craters around the planet, experts have worked out that almost the entire surface was renewed 500 million years ago. Some people think that Venus is like a giant pressure cooker. Deprived of the low-level, day-to-day volcanism found on Earth, it's unable to let off steam. Instead it gets more and more hot and bothered until, in the planetary equivalent of a blown gasket, it suddenly pops its cork, erupting lava from every orifice until its internal pressure is relieved.

Three types of Venusian volcano: a corona (collapsed volcanic bulge); an Earth-like volcanic cone, and a series of pancake domes made from congealed lava.

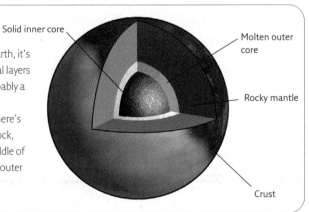

INSIDE VENUS

Since it's only slightly smaller than Earth, it's no great surprise that Venus's internal layers are quite similar to ours, though probably a bit further down the route to cooling completely. Underneath the crust, there's a churning mantle of semi-molten rock, then an iron and nickel core. The middle of the core has begun to solidify, but its outer parts are probably still molten.

Solid inner core

Molten outer core

Rocky mantle

Crust

Why no tectonics?

The big geological question about Venus is why its volcanism happens in sudden bursts rather than at a fairly steady rate. Venerologists (Venus experts) have known since the days of the earliest spaceprobes that the planet doesn't have a system of tectonic plates like Earth's. It's these plates, grinding past each other, pulling apart or pushing together, that drive much of Earth's volcanic activity. But this only raises a deeper question of why the plates aren't there.

There are several possible explanations. The simplest is that tectonics need a certain amount of internal heat to power them, and Venus's core could never quite cut it. Other ideas are connected to the lack of water (of which, more anon). Earth's plentiful liquid may help lubricate the 'aesthenosphere', the thin layer of runny material that the tectonic plates slip across. Water also helps to form lightweight carbonate minerals that give Earth's continents their different densities and determine which edge buckles up or is forced downward when plates push against one another. Venus's crust is probably nothing but dense, basaltic rock, so even if plates once tried to get moving, like a pair of obstinate rhinos they would have ground to a halt as they met head-on.

A view across the plains of Eistla Regio towards the great volcanoes of Sif Mons and Gula Mons.

Venera wreckage

While visiting Venus, you should be sure to check out one of the old probe landing sites. The Russians sent a whole fleet of Venera spacecraft here back in the early days of the Space Age, and seeing them in their current state is a salutary reminder of what the surface conditions can do (just in case you were getting comfortable inside that armoured suit).

Say what you like about the old Soviet Union, but they knew how to build things like tanks. Automobiles, aircraft and even architecture from the period have a certain innate tank-like quality to them, and the Soviet engineers soon extended this design philosophy to the problem of landing on Venus. After their first few probes lost contact shortly

Venera 14 lies gently corroding on the plains of Themis Regio. Although it stopped functioning within minutes of landing, the strength of the structure still holds it together even today.

after entering the atmosphere, sending back little more information than a few temperature and pressure measurements that rapidly disappeared off the scale, they got the hint that Venus was no ordinary planet. As a result, quite a few deteriorating relics of fine Soviet workmanship litter the surface of Venus.

The first colour view of the Venusian surface, as sent back by Venera 13. Not exactly paradise, is it?

Although Venera 7 was probably the first probe to make it to the surface intact (it sent back steady temperature and pressure readings for a few minutes before losing contact with Earth), the first pictures of the planet's surface came in 1975 from Veneras 9 and 10. The decayed remnants of Venera 9 still show collapsed parts of its landing cushion and the battered rim of its disc-shaped aerobrake. The heavily shielded main sphere is still surprisingly intact, but many of the metal struts and pipes that held it all together are corroded to the point of collapse.

WATER, WATER... NOWHERE?

One of the most surprising things about Venus is its total lack of water in any form. The region where Earth and Venus formed seems to have been the celestial equivalent of a monsoon zone, and clearly a lot of water ended up in or on Earth. So why didn't the same thing happen to Venus?

Perhaps it did, though scientists are still wrangling over the evidence or otherwise for ancient oceans on Venus. However, something must have happened to drive away the water. The best bet is that Venus's location slightly closer to the Sun meant that water evaporated into the atmosphere, where it was split by fierce ultraviolet radiation into hydrogen and oxygen. The lightweight hydrogen then blew away into space.

The neat thing about this theory is that it also explains why Venus has so much carbon dioxide in its atmosphere. On Earth, a reaction between water, carbon dioxide and the surface rocks (called chemical weathering) gradually turns the rocks into carbonate minerals and extracts carbon dioxide from the atmosphere. The lack of water, and therefore carbonates, probably also helped to throttle Venus's early attempts at plate tectonics (see 'Why no tectonics?', p.55).

If you can't stand the heat...

Mercury is an overlooked destination, mostly because it's so hard to reach as it races round the Sun. It's also a place of extremes, with searing daylight turning to frozen airless night in a matter of minutes. But if you can get there, it's a fascinating world. See the huge Caloris Basin, caused by a massive asteroid impact that sent shockwaves all round the planet. Climb the kilometre-high cliffs that separate different chunks of the planet's crust. And don't miss the fantastic double sunrises!

Venus may be the hottest planet in the Solar System, but with its permanent cloud cover, it's certainly not the place to top up your tan. Real sun worshippers are better advised to head for Mercury, the tiny and baking airless world that orbits closest to our star.

Although Mercury's surface isn't as reflective as those on some other planets, you'll still need tinted filters to find detail in the glare of the daylit side.

Cynics often dismiss Mercury as a nothing world, a coppery imitation of Earth's Moon at the sweaty end of the Solar System. It's true that Mercury doesn't have the variety of terrain found on the bigger worlds that have had more active histories, but it's still got some unique and spectacular features that are well worth seeing, and it makes an ideal base for exploring the neighbourhood of the Sun itself.

Getting there

The big challenge with Mercury is getting to it in the first place. Lying so close to the Sun that its year lasts just 88 Earth days means that, according to the immutable (and sometimes simply annoying) laws of planetary motion, it also travels faster in its orbit than any other planet, reaching a top speed of 48 km/s (30 miles per second). Okay, it's true that all motion is relative, and the Earth itself is no slouch with a nippy 30 km/s (19 miles per second) top speed, but that's still a heck of a speed difference to make up if you want

TRAPPED BY THE SUN

A mistimed rocket burn on the way to Mercury is the surest way to strand yourself in a permanent orbit close to the Sun. Moving at high speeds, you'll need a lot more fuel to boost yourself back into a nice long ellipse for return to Earth, and if you've not brought enough with you, it'll cost a small fortune to get a supply tanker to come and rescue you.

to join Mercury in its orbit, so you'll need a pretty powerful engine (or have to put up with a longer voyage and a lot of tedious mucking about with slingshot manoeuvres).

A lot of Mercury's best features are only apparent from orbit or on the surface. As you get closer you should at least be able to make out some of the planet's major craters with binoculars, which is more than you can do from Earth! Getting closer also puts more distance between Mercury and the Sun in your field of view. On Earth, Mercury always lies in the same part of the sky as the Sun, so you only ever see it in the twilight just after sunset or just before sunrise, never in a clear, dark sky. Pick out some features on the planet and chart their positions from day to day, and you'll discover another of Mercury's secrets: it rotates very slowly, spinning on its axis just once in every 59 days (precisely two thirds of a Mercury year). As we'll see later, this has bizarre consequences.

Once you reach orbit, you'll start to get a feel for what makes Mercury a bit different. The most obvious things are the *rupes* (say 'roo-pez'), huge faults that run across the cratered landscape. Then there's the Caloris Basin – hard to miss from a distance, but easily lost from low orbit on account of its sheer size.

If you know much about orbital mechanics, that'll give you another clue that Mercury is a bit weird. For a given distance from the planet, your orbit will be quite a bit faster than you might expect, and that's because the planet's mass and gravitational field are stronger than they should be if you compare Mercury to its planetary neighbours. You might also be surprised to find that Mercury's got a magnetic field – not bad for such a tiddler of a planet, especially when the much bigger Venus doesn't have one at all.

It's down on the surface, though, that you'll really notice the difference. Mercury's gravity is a lot stronger than the Moon's, and more than one third of Earth's. And you can't travel far without encountering one of the many *rupes* systems, features that look as if someone took a hammer to Mercury's surface when it was already fully formed.

MERCURY DATA

Good points:
Good for suntans.
Lots of potential stopovers

Bad points:
Tiny and hard to reach

Day length:
58.6 Earth days

Year length:
88 Earth days

Gravity:
0.38 g

Surface temperature:
-170 to 430 °C
-280 to 800 °F

Communications time:
5 minutes or more

Discovery Rupes

This 650-km (400-mile) long scar across Mercury's landscape is probably the most famous of the *rupes* features. In places, the cliff rises for two kilometres (1.2 miles) or more, and it splits several craters straight down the middle. It's as if one half of Earth's Manhattan Island was on a completely different level to the other, with an enormous vertical cliff running down the middle of Broadway. The view from the cliff edge is spectacular, but watch out for vertigo!

Discovery Rupes forms a precipitous cliff as it runs across a 50-km (30-mile) crater on Mercury.

Land or fly over the *rupes*, and you'll see one major difference from cliffs on Earth. On our planet, cliffs usually have strata, layers of different colour or texture that show how the rock formed from fine silt drifting to the bottom of a lake or sea and getting pressed together. Mercury's got no such sedimentary layers. Its cliffs are great blocks of volcanic rock lifted past each other.

These great slips in Mercury's landscape can also take other forms. In some areas, huge strips of land with cliffs on either side pop out of the surface. Elsewhere, other chunks of the landscape have collapsed inwards to form steep-sided depressions.

So just what did shake up Mercury's surface? It seems that the whole of the planet shrank at some point in its history, after the crust had solidified and most of the craters had formed. This probably comes down to Mercury's huge iron core, which continued to heat up for some time after the planet's formation, expanding and opening up cracks in the crust. When it eventually cooled and shrank, fragments of crust jostled against each other, some dropping down and others popping up like jigsaw pieces wrongly forced into a puzzle after someone's got frustrated filling in the sky.

LOW-GRAVITY DANGERS

Just because the gravity on Mercury and some of the other small worlds of the Solar System is weak, that's no excuse to get cocky. Fall off Discovery Rupes, for instance, and you'll keep accelerating all the way down. Unlike Earth, these smaller worlds have no atmospheric drag to slow you down, so you could end up hitting the ground at just the same speed as if you'd jumped out of a plane at a similar height on Earth. In a vacuum, the only 'terminal velocity' is the one that kills you.

Caloris Basin

Mercury's largest and most impressive feature is, without a doubt, the Caloris Basin. Second only in the Solar System to the South Pole-Aitken Basin on our own Moon, this vast impact crater is itself roughly 1,340 km (840 miles) across, and you can see evidence of its birth all over the planet.

The only problem, as with the Aitken Basin, is how to appreciate the crater's sheer scale. To get an overview of it, you need to be some distance from the planet itself, able to take in an entire hemisphere. Even then, you won't see it all unless Mercury's slow-moving daylight is illuminating the right area of the planet.

Shockwaves from the Caloris Basin impact spread around the planet to create the 'weird terrain' on the opposite side of Mercury.

The alternative is to fly in fast and low, and in many ways this is the better option. The Caloris Basin gouged such a hole into Mercury's crust that it triggered volcanic eruptions which allowed the basin floor to heal itself. As a result, the basin interior is fairly flat and featureless – quite like a lunar *mare*. But with no difference between the colour of the new crater floor and the surrounding rock, it's hard to make out the overall structure. Your best bet is to look for the rings of concentric mountain ranges and rays of hilly ejecta that surround the original impact site – and these are just as easy to spot from lower altitudes.

Huge concentric mountain ranges ring the central plain of the Caloris Basin.

Flying in towards Caloris at sunrise or sunset is an unforgettable experience. 1,800 kilometres (1,125 miles) out, you'll cross broad smooth plains with a long range of hills running across across them: the Odin formation. This is the outer limit of Caloris's ejecta. Closer in, around 1,000 km (600 miles) from the crater itself, you'll run through the Van Eyck formation – ranges of hills and grooves that spread out radially from the crater in every direction. These are the scars formed in Mercury's landscape as the ejecta sprayed outwards. Inward from this, you'll fly over the spectacular mountain ranges at the edge of the basin, and as you reach the crater itself, the shadows should also help you to pick out low undulations in the frozen lava blanket that now covers the interior of Caloris.

Including its ejecta blanket, the whole Caloris structure has a diameter of almost 5,000 km (3,000 miles), extending almost one third of the way around Mercury. But that's not the limit of Caloris's influence. Fly to the far end of the planet and you'll see more traces of the chaos caused by this huge impact.

When the 150-km (90-mile) Caloris meteorite hit Mercury, two sets of shockwaves went rippling out from the impact site. One lot radiated outwards through Mercury's solid crust, while the others spread downwards, through the planet's mantle and core, and eventually straight out the other side. At the point on Mercury exactly opposite the original impact, the two sets of waves met up again and threw a wild party to celebrate, leaving the metaphorical living room wrecked beyond repair.

The far side of the planet shook and trembled like a bowl of jelly perched atop the mother of all sound systems. Solid rock shattered into so much dust and boulders, and the wreckage bounced around until the waves finally began to subside. As the rocks, often semi-molten from the initial shock, began to settle back to the ground, they created a jumbled mass of hills, now known as Mercury's 'weird terrain', the planetary mapmaker's equivalent of 'Here be Dragons'.

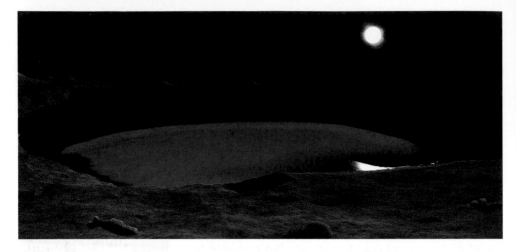

Polar ice

While on Mercury, it's worth taking in one of the most unusual sights in the Solar System. Bizarrely, a few select spots shelter frozen deposits of water ice, nestling right next door to rocks hot enough to melt lead.

These icy lakes are another of Mercury's similarities to our Moon. The angry Sun could zap them into so much evaporating gas without a second glance, but they escape its notice by sheltering in deep craters close to Mercury's poles. Because Mercury orbits the Sun almost bolt upright, there are parts of the surface where the Sun never rises more than a few degrees above the horizon, and where the floors of some deep craters never see a sunrise. Here, ice deposits dumped by comet impacts can remain in pristine condition for billions of years. The chances of a comet hitting the sweet spot may seem slim, but, over the life of the Solar System, it's virtually bound to happen not just once, but many times over.

The sight of baking rock and ice directly neighbouring each other is a strange one, but with no atmosphere to spread heat from one to the other, the two seem to get along just fine.

Light from the Sun's corona is enough to illuminate an ice-filled crater at Mercury's south pole without evaporating its precious contents.

LOOKING AT THE SUN

At the risk of nagging, we'd like to emphasise again how important it is not to look directly at the Sun. Even on Earth, staring directly at our local star is enough to permanently damage your retina. Further in, you can be blinded in an instant by a foolish glance in the wrong direction. All those visors and shades on your helmet are there for a purpose, so use them!

The double sunrise

Not so much a place, this, as an event, and one that's well worth catching if you're on Mercury near perihelion, as the planet comes closest to the Sun. Pick the right spot at the right time, and make sure you have a video camera handy.

What's all the fuss about? Well, some parts of Mercury don't just get to see one sunrise and sunset per day – they get two! It's a mind-boggling idea, so to appreciate what's going on, let's take it step by step. Most planets, including Earth and Mercury, spin so that their surface moves from west to east. In the course of one day, this usually means that the Sun will rise in the east and set in the west. Because the planet is orbiting the Sun, though, the Sun's position will drift from west to east, completing one circuit in a year, normally moving too slowly to notice in the space of a day.

On fast-moving, slow-rotating Mercury, though, years are short and days are long. The Sun drifts a long way from west to east in a Mercurian day, loitering in the sky like a teenager in a shopping mall before finally being dragged away by the planet's rotation. The combination of motions means that the time from sunrise to sunrise is actually three Mercury days or two Mercury years long, equivalent to 176 Earth days.

So far, so weird. But Mercury also has an elongated orbit, so it moves quite a bit faster along its orbit at perihelion than at aphelion. Right around perihelion, the west-east movement caused by Mercury orbiting the Sun can actually outpace the east-west drift from the planet's rotation, causing the Sun to falter in the sky and drift backwards for a few Earth days. If you're in the right place at the right time, on the borderline between Mercury's long night and day, you should be treated to a weird sight. The Sun peeks above the horizon, then apparently decides it can't be bothered, slinking back to bed before finally thinking better of it and hauling itself back up to begin its long crawl across the sky. On the other side of the planet, meanwhile, the Sun sets and then comes back for an encore.

A sequence of time-lapse images captures the advance and retreat of sunrise on Mercury.

Further afield –
Vulcanoids

This is one of those parts of the guide where we're future-proofing ourselves. We don't want to get served with a writ several centuries from now, saying 'I was just crossing the space lanes from Mercury to the Sun, keeping a careful eye out for comets cutting across my path, when BAM! Suddenly this rock comes straight at me. And you guys didn't say anything about it, so pay up!'

So, for the benefit of litigious future travellers, here's a little about the vulcanoids (and yes, we know they sound like a bad sci-fi alien – get over it). So far, these small rocky worlds are entirely hypothetical, but there's no reason why they shouldn't exist. In theory, there's a stable region of the Solar System inward of Mercury where small objects could have formed in circular orbits very close to the Sun and might still lurk to this day. These worlds would be made almost entirely of the least volatile metals – this close in, most elements will just evaporate in the intense heat and then be blown away on the solar wind.

Astronomers have been looking for vulcanoids since the late twentieth century, when they finally got over their disappointment that the planet Vulcan itself didn't exist (see 'Vulcan', opposite). The most popular way to search is by photographing the sky around the Sun during a total eclipse, but since eclipses are thin on the ground, one alternative is to take an aeroplane up to the edge of the atmosphere and photograph the horizon straight after sunset. On the Moon, of course, night comes on pretty much instantly, so you don't have the glare of sunset to contend with, but we've still drawn a blank so far. If they're out there, they must be very small – certainly no bigger than a few kilometres across – but we'll probably never be able to rule them out for good.

In theory, the vulcanoids could occupy a narrow band around the Sun, inward of Mercury.

VULCAN

The original Vulcan (not the one *Star Trek*'s Mr Spock comes from) was a small hypothetical planet that hovered around the edges of respectability for much of the nineteenth and early twentieth century. Its existence was first proposed by famously argumentative French mathematician Urbain Le Verrier. Flushed with success after his mathematical discovery of Neptune (see p.181), Le Verrier was approached in 1859 by a French amateur astronomer who claimed to have seen a small object crossing the face of the Sun. Pretty soon, Le Verrier had come up with a prediction for a planet that matched the observations, and could also account for unexplained wobbles in Mercury's orbit. As is so often the way, as soon as Le Verrier announced his planet, people started 'seeing' it.

That's not to say the whole thing was a delusion. A lot of very well qualified people definitely saw something (or things). But a lot of other people who were looking at exactly the same time (for example, during solar eclipses) saw nothing. The whole saga dragged on with claim and counterclaim for pretty much the rest of the century, until in 1915 a young clerk from the Swiss Patent Office by the name of Albert Einstein came up with a better explanation for Mercury's orbital wobbles – the General Theory of Relativity.

So what did all those people see? Probably a mix of phenomena, such as otherwise unnoticed sungrazing comets, unrecorded Near Earth Asteroids or other bits of space junk, and perhaps even the occasional genuine vulcanoid. But as for Vulcan itself, it's probably best left to the Trekkies...

Man conquers the vulcanoids! Perhaps one day we'll actually find one of these elusive rocks – but if we want to claim it, the flag had better be fire-retardant!

Further afield –
Close to the Sun

An enormous solar flare looms over the mottled limb of the Sun in this ultraviolet image. Note the flame-like spicules rising from the visible surface.

Mercury is probably as close as most people will ever want to get to the Sun, but of course we can't go this far without at least talking about the possibilities for exploring our local star up close and personal.

To put things simply, the Sun is a ball of exploding hydrogen gas. Its surprisingly neat, spherical surface is roughly 1.4 million kilometres (865,000 miles) across, but its outer atmosphere (called the corona) stretches well beyond that, reaching several million kilometres into space in constantly shifting lobes of hot, transparent gas. Ultimately, the corona blends into the solar wind – the stream of high-energy particles blowing out across the Solar System in every direction. This is the stuff that gets caught up in the magnetic fields around planets like Earth and causes aurorae, or northern and southern lights.

Sunspots come and go constantly from the surface, occasionally growing into larger complex groups like this one.

Although Mercury itself is quite a bit closer to the Sun than Earth, it's still too far away for you to see the Sun's surface in detail. You'll also need even thicker and more absorbent filters than those required to watch the Sun safely from Earth, so you might as well resign yourself to looking at it through a properly filtered telescope. That means one with a great big safety-certified opaque filter on the front, NOT a weedy piece of darkened glass over the eyepiece! On Earth, you can also project the Sun's image through the telescope onto a piece of card, but that doesn't really work this close in. For one thing, the image of the disk gets so brilliant that it washes out any darker features, and for another, there's a much bigger risk of setting your piece of card on fire, which is not such a good idea in the sealed atmosphere of a spaceship.

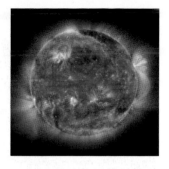

This false-coloured ultraviolet image reveals the magnetic turmoil in and around the Sun's surface.

INSIDE THE SUN

The Sun's interior is split into three main zones. The core is where the action is, as fusion reactions weld the nuclei of hydrogen atoms together to make the next lightest element, helium, and this elemental alchemy releases the energy that powers the Sun. Next comes the radiative zone. Energy shoots out from the core as high-energy gamma rays, but the material in this zone is so dense that the rays can't travel in a straight line. Instead they bounce around, gradually transferring energy to the gas atoms around them, and losing strength themselves. If you could trace the path of a single gamma ray, you might find that it took up to 100,000 years to make it across the radiative zone.

Convective zone

Radiative zone

Core

Photosphere

The radiative zone ends where the convective zone begins. Here, chemical changes inside the Sun mean that its gases become opaque. Radiation can't cross the boundary, so it gets absorbed into the gas at the bottom of the zone, heating it up and causing it to rise through its surroundings. Finally, at the photosphere, the Sun's 'outer surface', where temperatures drop to 5,500°C (9,900°F), the gas becomes transparent again. Energy from the top of the convection cells can now escape as a variety of radiations including visible light, but this time the Sun's density is low enough that the rays can just shoot off in straight lines, carrying heat and light to the worlds of the Solar System.

Even if you can get a good look at the Sun, what are you going to see? Probably not a lot without the correct filtering, to be honest. The Sun throws out so much radiation that detail easily gets drowned out, and usually the only obvious features are dark sunspots, Earth-sized regions of the solar surface where temperatures fall away to a relatively cool 3,500°C (6,300°F), so they show up dark in comparison to the rest of the photosphere (see 'Inside the Sun', p.71) at 5,500°C (9,900°F).

To really get a feel for activity on the Sun, you'll need to use a filter that blocks out all but a narrow band of wavelengths, or detectors that image the Sun at radiations other than visible light. Hydrogen alpha filters are a popular choice. They block out all of the Sun's radiation except for a specific wavelength produced by hydrogen in the atmosphere. Although this gives everything a rather orangey glow, it reveals a huge amount of detail in the photosphere and other upper layers of the Sun. If you plan on going anywhere much closer to the Sun than Mercury, you'll need one to help navigate the dangerous but invisible hazards of the solar neighbourhood.

With filters in place, the Sun's brilliant disk is transformed into a seething, constantly changing bauble of dark and light regions. The entire surface is broken up into a pattern of dark-edged cells called granulation. Each granule is the top of an individual convection cell, where rising gas releases its heat in the centre, cools, and sinks down around the edges. Bright structures called filaments snake above the granulation – these are areas of the solar atmosphere where the gas is denser and hotter. Get close enough, and you should also be able to see the spicules, pillars of flame, up to 10,000 km (6,000 miles) long, that tower above the photosphere and carry hot gas into the upper atmosphere. People often forget that the upper transparent layer of the Sun's atmosphere, the corona, is much hotter than the photosphere, with temperatures reaching 1 million °C (1.8 million °F) or more.

At solar minimum (top), the Sun's surface looks relatively calm, broken up into dark-edged convection granules. At maximum (above) the surface is in turmoil with numerous bright areas associated with magnetic activity, and huge flares looping through the atmosphere.

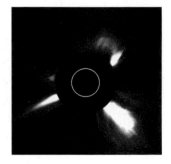

A coronagraph is simply a disk that blocks out the direct image of the Sun's disk, revealing details in the atmosphere – in this case three huge eruptions of hot gas called 'Coronal Mass Ejections'.

Above the photosphere, lots of other features come and go, including small loops of cool, reddish gas called prominences, and much more spectacular solar flares – enormous eruptions of gas that escape the Sun entirely and flood the inner Solar System with high-energy solar particles.

Filters that detect radiations such as ultraviolet light will show an even more spectacular difference in the Sun. Ultraviolet radiation comes from the hottest regions of gas, and it reveals a complex structure of loops and swirls in the solar surface. Switch between one view and another, and you should be able to see how ultraviolet features coincide with visible light structures such as sunspots and promi-nences. The tangled ultraviolet view, though, makes it a lot clearer that gas on the Sun's surface is following shifting curves, loops and lines. It's a visible manifestation of the solar magnetic field, the real driving force behind the Sun's activity (see 'Solar cycle', right).

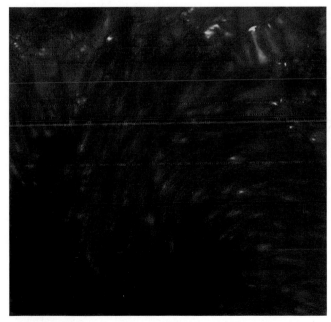

An extreme close-up of a sunspot's edge reveals the Sun's turbulent surface. Each of the thread-like spicules is several thousand kilometres long.

SOLAR CYCLE

Solar activity is a constant hazard in the inner Solar System, but you can keep the risks to a minimum by tracking the 11-year solar cycle. With occasional exceptions that no one really understands, the amount of solar activity – sunspots, prominences and flares – rises and falls in a regular pattern.

The cycle is linked to the Sun's magnetic field. Because it's not trapped in a solid body like the Earth, this field changes constantly. It is generated beneath the Sun's surface by currents running through electrically charged gas, but as different bits of the Sun, spinning at different rates, wind themselves around each other, the field becomes twisted and starts to bulge out through the surface in the solar equivalent of a hernia. Where loops of field emerge and re-enter the Sun, sunspot groups form. Sometimes, the loops short-circuit, releasing enormous amounts of energy and creating flares. Eventually, sunspot loops start cancelling each other out, destroying the magnetic field, which is then regenerated from scratch.

Further afield – Comet surfing

Every couple of months, on average, the inner Solar System gets a new visitor – a ball of rock and ice that comes plunging towards the Sun, usually fresh from the planetary deep freeze beyond Neptune. Get your timing right, and you can hitch a ride around the Sun with one of these flying visitors. Many of those who've done it say it's the trip of a lifetime.

If you can catch your comet somewhere around the asteroid belt, you should get the chance to watch it come to life. That far out from the Sun, the average comet nucleus will look like a very dark, potato-shaped lump, probably a few kilometres across (unless you've caught a whopper).

This is a great time to pop down to the surface and have a poke around before things get too dangerous. Make sure you anchor yourself to the ground as soon as you land,

though: most comets have such weak gravity that one wrong move could send you spinning off into space.

Take a trowel to scrape away at the surface, and you'll find that the comet's darkness is only skin-deep. Just a few centimetres down, the outer crust gives way to dirty, but unmistakably icy, material. It's mostly water ice, with small amounts of other frozen volatile chemicals mixed in – methane, ammonia and carbons monoxide and dioxide, for example. The surface, meanwhile, is a mix of dark dust grains – the stuff that's left behind when the ice has evaporated. That too is often rich in carbon-based 'organic' chemicals.

If you're lucky, you may be able to find early signs that the comet is waking up. As its surface rotates, watch where the newly risen Sun strikes it and look for puffs of steam escaping into space. The comet's dark surface makes it a good absorber of heat from the Sun, and as the ice beneath the crust gradually gets the benefit of this heat,

COMET CATCHING

Early probes to comets were usually targeted at the predictable short-period variety, but in these days of early-warning outposts and long-range telescopes, we get more notice of approaching long-period comets. For the full comet experience, it's worth trying to hitch up with one of these, perhaps around the orbit of Mars or Jupiter.

The surface of a comet is an eerie site as it wakes from its long slumber and jets of high-pressure vapour shoot from beneath the dark crust.

pockets can evaporate, cracking the surface apart and blasting off into space. Despite the comet's weak gravity, much of the escaping gas remains trapped in an extended 'atmosphere' around the nucleus, gradually building up to create a planet-sized halo or 'coma', with the nucleus buried at its centre.

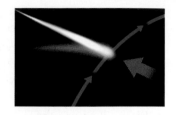

Once around the Sun

A comet's blue-white gas tail is always blown straight away from the Sun by the solar wind (yellow arrow). The heavier particles of the yellowish dust tail tend to bend towards the path of the comet's orbit (blue arrow).

By the time you're past the orbit of Mars, your comet should be well and truly awake. As heat from the Sun increases, more cracks open up in the surface, and once the crust has warmed through, they no longer fall dormant again when the comet's rotation carries them back into darkness. The coma grows denser until your entire sky is filled with a milky-white haze, the Sun and stars dimly visible through it.

But you may start to notice something else. As the strength of the solar wind increases, it begins to affect the gas jets, bending them so they trail away from the Sun. Congratulations – your comet's got a tail.

Comets often disintegrate as they pass around the Sun – just look at the state of Comet Linear of 2001.

Pull away now to fly alongside the comet at a greater distance. You've seen the active nucleus, and things can get dangerous if you're too close as the comet rounds the Sun. Also, from outside the coma, you'll be able to truly appreciate the comet's scale. The small iceball at the centre may now be surrounded by a coma larger than Jupiter, with a tail extending for millions of kilometres behind it. In fact, there's usually more than one tail. First, there's a bluish-white ion tail of gas, electrically charged through its meeting with the solar wind, and always pointing directly away from the Sun. Second, there's a yellowish-white dust tail, blown in the same general direction but, because its particles have more inertia, curved back along the comet's orbital path. Both tails are at their longest and brightest as they come close to the Sun, but this is also the time of maximum danger for the comet. Its internal structure may be fatally weakened by the blast of solar heat and the gravitational stresses of a close approach to the Sun. The doomsday scenario sees the comet disintegrate completely, its individual chunks plunging into the Sun or evaporating like a snowball in a heatwave. Even if the comet survives, mountain-sized slabs of ice may detach themselves and drift away from the original nucleus. This is why a comet is dangerous to be around during perihelion.

As your comet starts to retreat from the Sun, the tails will now lead the way, blown ahead of the comet's path by the solar wind. Its activity may already be dwindling, so now is the best time to wave it farewell, and head for home.

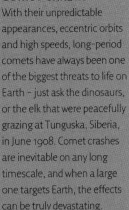

COMET CRASH!

With their unpredictable appearances, eccentric orbits and high speeds, long-period comets have always been one of the biggest threats to life on Earth – just ask the dinosaurs, or the elk that were peacefully grazing at Tunguska, Siberia, in June 1908. Comet crashes are inevitable on any long timescale, and when a large one targets Earth, the effects can be truly devastating.

Fortunately for us, Earth has a couple of guardian angels. The first is Jupiter, whose gravity is so powerful that it frequently disrupts comets, breaking them apart and pulling them to their doom, or at least affecting their orbits, turning them into more sedate short-period comets. Good news for us, but potentially bad news for any life trying to evolve on the Galilean moons, which are directly in the firing line. Our second guardian is our own Moon, which has also soaked up more than its fair share of large objects that might otherwise have played havoc with Earth.

As a comet's nucleus rounds the Sun, it will have enfolded itself in a halo of gas and dust – its coma.

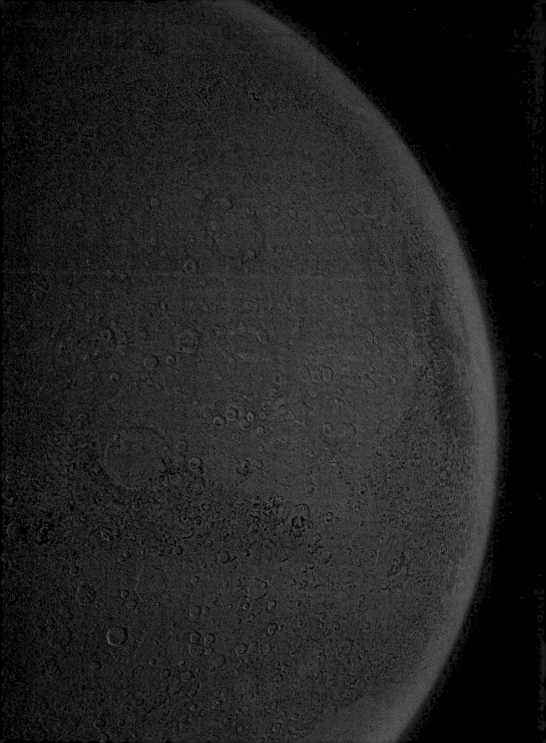

A warm welcome on the Red Planet

Mars combines ease of access, a relatively hospitable climate, spectacular scenery, and more than a hint of mystery and romance – small wonder it's such a popular destination! Whether you choose to go skating on the polar ice caps, abseiling in the Valles Marineris canyon, climbing on the towering volcanoes or fossil-hunting in the dried-up ocean basins, the Red Planet offers something for everyone.

On any tour of the Solar System, Mars is one of the highlights. Despite its small size, this little red world offers a huge variety of scenery and constant surprises. The time delay on a call home to Earth will vary wildly depending on when you plan your visit. When Mars and Earth are at their closest approach, signals take just three minutes to travel between the worlds. At greatest separation, they take around 22 minutes, so conversations with your loved ones may get a little stilted. The difference is so great because Mars has quite an eccentric orbit – its distance from the Sun ranges from 205 to 250 million kilometres (128 to 156 million miles).

Thanks to its eccentric orbit, Earth's closest approach to Mars can range from 57 million km (35 million miles) to 99 million km (61 million miles).

Getting there

Martian weather is very seasonal and the timing of your visit could have a major effect on how much you get out of it. The climate is naturally on the chilly side – at an average of 70 million km (43 million miles) further from the Sun than Earth, typical daily temperatures are around -55 °C (-67 °F) – you'd be well advised to pack your thermal underwear. However, Mars is tilted on its axis at a similar angle to Earth (25 degrees instead of 23.5) and rotates once every 24 hours 40 minutes, so the pattern of days and seasons is pretty similar to Earth. Just bear in mind that the seasons tend to last that little bit longer – a Martian year is 780 Earth days long.

The orbit complicates matters for Martian meteorologists. The big difference between perihelion and aphelion (closest and furthest points from the Sun) affects the amount of solar heating Mars receives, and this can tend to exaggerate or soften the seasons. Most importantly, try to resist the temptation offered by fast and cheap travel during Martian perihelion, as this tends to be peak storm season, with huge clouds of fine red dust whipped up in the atmosphere. While you should be safe inside a well-made spacesuit, the dust will certainly spoil some of the spectacular views.

Winter sports enthusiasts planning to visit the planet's polar ice caps will also want to check conditions in advance. The caps are largely made of solid carbon dioxide (so-called dry ice) and tend to evaporate into the atmosphere during the spring, settling back into new layers of frost at the other pole as it moves into its autumn. This gives the Martian polar caps a spectacular and beautiful layered appearance, but it also makes the ice a moving target – check with your travel agent before departure, so you don't end up at the wrong pole! In general, however, the ice is thicker and more reliable at the north pole, where a cap of permanent water ice underlies the carbon dioxide.

MARS DATA

Good points:
The planet most like Earth. Spectacular landscapes

Bad points:
Dust storms. No Martians

Day length:
24 hours 37 min

Year length:
1.88 Earth years

Gravity:
0.38 g

Surface temperature:
-130 to 30 °C
-210 to 80 °F

Communications time:
Shortest: 3 min 8 s
Longest: 22 min 13 s

The view from orbit

A brief stopover in orbit, a few hundred kilometres above the Martian surface, is an ideal opportunity to get an overview of the entire planet. Assuming you've avoided the storm season, the difference between the northern and southern hemispheres is unmistakeable. The south is mostly heavily cratered highlands, broken by the huge sandy basin called Hellas. The north, by contrast, is dominated by low rolling plains, winding valleys, and, of course, Tharsis.

The Tharsis Bulge is Mars' pot belly. Sharp-eyed travellers can sometimes spot that something's wrong with the planet's shape from a few million kilometres out, but the bulge is so huge that it's impossible to see from orbit, 5,000 km (3,000 miles) across, and an average of 10 km (6 miles) above its surroundings. Its major attractions are unmistakable, however – the chain of three volcanoes along its north-eastern side, the huge dome of Olympus Mons, the Solar System's largest volcano, and the great gash of the Mariner Valleys along the bulge's southern edge.

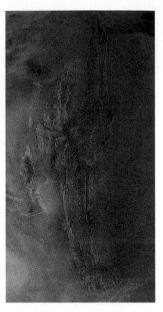

Only photos from high orbit (above) can take in the full extend of the Valles Marineris complex. Even from the edge of the atmosphere (below) you'll only see a small area.

Martian canals

The curious story of the canals of Mars began when Italian astronomer Giovanni Schiaparelli fell victim to an optical illusion that some people still see when studying Mars through a telescope today. The human eye loves to find patterns in randomness, and has a natural tendency to 'join the dots'. In this case, the dots were the darker features on the Martian surface, and Schiaparelli believed he could trace straight lines between them. He published his unusual observation in 1877, using the word *canali* to describe the features. In Italian, *canali* just means 'channels' – with no implication of design or intelligence. In English, of course, it's a different matter, and the world leapt on Schiaparelli's discovery. Soon everyone was seeing the 'canals', and speculating on the creatures who had built them, their purpose, and their layout. One popular idea was that they were irrigation channels, carrying water from the icy Martian poles to plantations around the arid equator. Several maps were published in reputable scientific journals before the truth began to emerge. Many experienced astronomers could not see anything, even with guidance from those who could. The quality of the instrument didn't seem to be important, so it wasn't a case of new technology leading to a new discovery. The only explanation was that the canals were an illusion. Far from being the work of intelligent life, they never existed in the first place.

Valles Marineris

The Mariner Valleys, named after the plucky Mariner 9 spaceprobe that discovered them, are a must-see for any Martian tourist. A trek on foot to the edge of the canyon is not for the faint-hearted – on a clear day you can see to the very base of this vast fissure in the planet's crust, 10 km (6 miles) below. Be sure to check a map beforehand – in many places the canyon walls are unstable. Weakened by daily temperature changes and blasted by dust storms, they could collapse beneath you even in the weak Martian gravity. If you do make it to the edge, you probably won't be able to see the other side – individual valleys can be as much as 100 km (60 miles) wide.

To really appreciate the scale of the Valles Marineris, take to the air. Martian aircraft need a huge wingspan to give them lift in the thin atmosphere, but once they're up, they can keep flying for ages, usually requiring nothing more than solar energy. From a few kilometres up, you can see how Mars's 'Grand Canyon' dwarfs the one back on Earth. At their widest, the interlinking valleys create a gouge in the Martian surface up to 600 km (375 miles) wide. Early morning flights can be particularly stunning, as the rising Sun burns away overnight mist that often forms in the valley.

Unlike the Grand Canyon, the Valles never had a river running through them – they seem to be a huge crack in the Martian crust, created as it buckled under the weight of the Tharsis Bulge, which rises to the north of here. Some astronomers think they were created by huge mudslides billions of years ago. Heat from the Tharsis volcanoes could have melted ice just below the surface, causing it to burst out in catastrophic flash floods, liquefying the ground and giving this region of Mars a serious case of subsidence.

Although there was never a single river carving its way through the channel, water naturally collected here after the valleys had formed, and you can still visit the dried-up lake beds on the canyon floor. Many of these are covered by dark drifts of fine Martian soil, which can be stirred up by high

MARINER 9

Back in the 1970s, the first spacecraft to enter orbit around Mars was NASA's Mariner 9. Before that, three other probes had flown past the Red Planet, but none carried braking rockets that would allow them to slow down, so they sent back just brief snapshots of the surface. Disappointed astronomers, some of whom had spent much of the previous century speculating wildly about the prospects for life, saw these early pictures of the planet's cratered highlands, and reluctantly concluded that Mars was a dead, moonlike world. When Mariner 9 reached Mars during its perihelion in November 1971, it arrived at the height of the storm season. The mission scientists had to put up with several weeks of pictures showing nothing but dust and clouds, before the atmosphere finally cleared, giving them their first glimpse of the northern hemisphere's valleys and volcanoes.

winds blowing along the canyon even outside the storm season. In places the drifts pile up to form huge dune fields.

Experienced abseilers can also find colourful layered rock strata on some of the cliffs – sedimentary rocks that reveal these areas were once underwater. At their western end, the Valles die out in the mazelike channels of the Noctis Labyrinthus, the romantically named Labyrinth of Night. These channels seem to have been formed by the same process as the main canyon system, but for much of their history they could have acted as drainage channels, carrying water from the high surrounding plains into the canyon.

From ground level, the Valles Marineris are split into a steep ridges with broad plains between them. The scale of the entire complex is only clear from orbit.

INSIDE MARS

Being a lot smaller than the Earth, Mars has cooled down from its formation a lot more quickly. It seems to have a smaller and less dense core than Earth, maybe with lighter elements mixed into the iron. Whatever its composition, the core has probably solidified now, but it might still be hot enough to keep the mantle warm. The crust varies a lot across the planet. Under the southern highlands it's up to 80 km (50 miles) thick, while beneath the northern plains it's just 35 km (22 miles) deep.

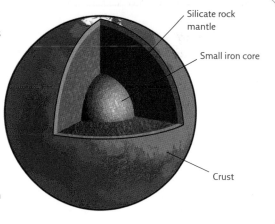

Silicate rock mantle

Small iron core

Crust

Olympus Mons

The vast calderas at the peak of Olympus Mons are best appreciated during a dawn flight.

After the Valles Marineris, the other standout Martian attractions are the great volcanoes of the Tharsis Bulge. There are four in total. The three smaller ones (Arsia, Pavonis and Ascreus Montes) form a chain running south-west to northeast across the bulge. Each of these is a monster in Earthbound terms, rising higher above the Tharsis Bulge than the Hawaiian volcanoes rise from the floor of the Pacific ocean.

The fourth stands apart, to the southeast of the chain. This is the famous Olympus Mons, the largest volcano in the Solar System.

First-time visitors often express disappointment with Olympus Mons, and it's true that it's hard to appreciate the whole thing from ground level. Then again, that's no wonder considering it's more than 500 km (300 miles) across, and rises to its 27-km (17-mile) peak on a slope that's barely perceptible to the average Martian pedestrian. The best way to get an idea of the overall scale of Olympus Mons is from a low orbit, where you're far enough up to take in the whole thing. Keen photographers should try to

Notice the clouds hugging the peaks of the Tharsis volcano chain? Like mountains on Earth, the Martian volcanoes affect the local weather.

fly over around local sunrise or sunset, when the long shadows will exaggerate the relief and reveal the rippling lava flows along the flanks.

Of course, part of the disappointment lies in the fact that Olympus Mons doesn't look like Mount Saint Helen's or Vesuvius, but there's not a lot we can do about that. It's a different type of volcano entirely, formed by the slow build-up of layer upon layer of lava welling up through a great crack in the Martian surface over millions of years. Even when it was erupting, Olympus Mons probably wouldn't have been the spectacular sight you'd hope for – it's unlikely that it was ever as explosive as smaller volcanoes on Earth. Of course, the question of whether Olympus Mons or any of the other Martian volcanoes will ever erupt again is still hotly debated by planetary scientists (see 'Mars erupts!').

The most spectacular views are to be had from the cliffs surrounding the central caldera complex. Overall, the caldera is 90 kilometres (56 miles) wide, ringed by sheer cliffs up to 6 km (3.7 miles) high. The complex looks like a series of overlapping impact craters, with floors at different levels. The largest area, surrounded by ring-shaped faults, was active around 140 million years ago.

MARS ERUPTS!

Are the Martian volcanoes all as dead as they appear? According to most ideas about planetary geology, they should be. Mars's relatively small size means that its core ought to have cooled down rapidly and its interior would have long since frozen solid. Mars itself, however, obviously knows better than the planetary geologists, because it flaunts the signs of recent volcanic activity for all to see. True, the eruptions may not have been on the same scale as the Tharsis volcanoes, but the pristine condition of some lava flows, unscarred by even a peppering of impact craters, can't be argued with. According to the latest ideas, some parts of Mars were volcanically active as recently as a few million years ago. That's just the blink of a geological eye and so most geologists, more or less grudgingly, admit that Mars today is probably still volcanically active, even if they don't know why!

The huge Olympus Mons caldera contains a number of craters that were active at different times. The big central structure with concentric faults around it is the most recently plugged-up fissure.

The Beagle has landed (with a thump!)

This image from the Beagle recovery team shows the wreckage of the doomed probe scattered across the Martian surface. Note the twisters in the background.

The sands of Mars, and indeed the history of early Martian exploration, are strewn with the wrecks of failed spacecraft. For a while back in the early days, it seemed that one in every two Martian spaceprobes went wrong for one reason or another, and even the most hard-headed technicians began to wonder about a 'Martian curse'.

Amongst these failures, one of the standouts must be Beagle 2, if only because it was such an unlikely project in the first place. Designed and developed by a team led by an eccentric British scientist (is there any other kind?) called Professor Pillinger, Beagle 2 was a tiny lander designed to hitch a lift on a larger European spaceprobe called Mars Express. It was the first probe sent to Mars specifically to look for life, and, coming at a time when the British space programme was virtually non-existent, it had to go out of its way to look for publicity and funding. As a result, Beagle 2 was probably the trendiest space probe ever, with a colour calibration system designed by controversial Brit artist Damien Hirst and a radio test signal composed by classical rock band Blur.

Sadly, despite years of effort and substantial amounts of money, something went wrong, as it often does in

LAWS OF SALVAGE

With such a long list of spaceprobes lost on or around Mars, it's no surprise that we've still not tracked them all down. If you stumble across anything artificial-looking during your travels on Mars, we recommend that you keep your distance and contact the appropriate authorities. Some probes still have a bounty for their recovery, and even if you don't find one of those, you'll still get a footnote in the history books.

spaceflight. Beagle separated from Mars Express flawlessly and appeared to make a textbook entry into the Martian atmosphere, but the signal from Mars never came. When orbiting survey spacecraft later tracked down the remains, it seemed that Beagle was probably just unlucky – it may have smashed straight into a hillside instead of landing on a reasonably flat piece of ground, and that could have stopped it deploying the airbags that were supposed to cushion its landing

Mars Polar Lander as it should have looked after landing on the Martian polar cap.

Although the wreckage was cleared away by an investigation team shortly after the first manned landings, Beagle's landing site is still well worth visiting. The team deliberately picked a region where they thought life might have found a toehold in the warmer and wetter Martian past. Isidis Planitia is the third largest impact basin on Mars, and lies just north of the equator at the boundary between the highlands and the lowland plains. This is where Martian oceans occupying the northern half of the planet might have lapped against the southern shores.

A CATALOGUE OF FAILURES

Mars 1: The first Soviet attempt at a Mars probe, lost in Earth orbit, 1962.

Mariner 3: An early NASA attempt, lost in Earth orbit, 1964. However, its twin Mariner 4 made the first flyby of Mars in July 1965.

Mars 2: Lost in a dust storm while attempting a soft landing, 1971. Mars 3 made it to the surface on 2 December 1971, but lost contact 20 seconds after landing.

Mars 4: A Soviet orbiter, this missed the planet entirely in 1973, as did the Mars 6 lander. Mars 7 was also lost, somewhere in the atmosphere. Only the Mars 5 orbiter made it safely to its destination.

Phobos 1: A 1988 Soviet attempt to investigate the Martian moons came to an embarrassing end when someone accidentally told the stabiliser systems to switch off in mid-flight. Phobos 2 made it to Mars, but lost contact in Martian orbit.

Mars Observer: NASA's 1992 attempt to return to Mars came to an end when its fuel tanks exploded as it entered Martian orbit.

Mars Polar Lander: A 1999 attempt to reach the ice caps went 'splat' when its retro-rockets shut down during descent.

Mars Climate Orbiter: Another 1999 probe, lost as it entered orbit. It turned out that some of the onboard systems were programmed to use metric units, while others were designed to use old-fashioned imperial units. Doh!

A tale of two craters

Next stops on our tour of the Red Planet are the landing sites of two more early spaceprobes, both rather more successful than the unfortunate Beagle. The Mars Exploration Rovers Spirit and Opportunity (NASA always loved to give its space probes the kind of names you'd expect to see on a luxury yacht) were a pair of robot vehicles that landed on opposite sides of the planet in 2004. Their sites are both worth visiting because this is where they found the clinching evidence that Mars was once much warmer and wetter than it is now.

An artist's impression of Mars as it might have looked back when the northern plains were covered with shallow oceans.

Gusev Crater

The Spirit rover came down more or less in the centre of this large crater, 170 km (105 miles) across. Gusev lies at the northern end of the Ma'adim Vallis, a winding channel that was clearly carved out by water over million of years, back in the ancient Martian past. It certainly looks as though the valley was draining water out of the surrounding highlands and into the crater. NASA's boffins were quietly confident that the crater floor would turn out to be covered in sedimentary rocks, formed as eroded soil carried in the river settled onto the lake bed and got compressed.

A view across Gusev Crater, landing site of the Spirit rover. Ma'adim Vallis, which once drained into the crater, can be seen leading out of the south rim into the distance.

Well, that was the theory, but things turned out to be a bit more complicated. The crater floor is actually covered in a deep layer of bog-standard Martian volcanic sand. There are plenty of rocks scattered on the surface, but those, too, are volcanic, with none of the tell-tale signs that they were ever underwater. Even the nearby Columbia Hills, which everyone hoped might be an outcrop of sedimentary rock, turned out to be depressingly volcanic. The only sign of water activity at all was a rock nicknamed Pot of Gold, covered in nodules of various minerals including iron-based mineral haematite, which on Earth usually forms when there's water around.

So why no water? The best guess is that a lot of Gusev got resurfaced with volcanic material after the water disappeared. And there's a likely culprit for this – the huge Apollinaris Patera volcano just a couple of hundred kilometres to the north.

Merdiani Planum

Fortunately, Opportunity found a different story on the other side of the planet. Its landing site was a relatively small plain that might once have been on the coast of an ancient ocean, roughly the size of Earth's Baltic Sea. The rocks here turned out to contain plenty of haematite, a good clue to the presence of water. More impressively, Opportunity was within striking distance of a small crater called Endurance.

For today's visitor, Endurance Crater offers a glimpse of the Martian interior. It's just 130 metres (420 ft) across, probably about 3–4 billion years old, and shallow enough to walk into. The miniature cliffs around its edges reveal layered sedimentary rocks, a dead giveaway that the rocks where this crater formed were themselves created underwater.

Small dark pebbles scattered across the surface are nicknamed 'blueberries'. They are rich in haematite, and probably formed elsewhere in the shallow sea. At the centre of the crater is a beautiful little dune field – you can still trace the tracks where Opportunity trundled down to its edge, before the NASA scientists thought better of risking their precious rover.

Sedimentary rocks lie exposed around the rim of Endurance Crater.

Opportunity took this photograph before it began its descent into Endurance Crater. Note the miniature dune field at the centre.

The Face and Cydonia

Ask any Martian tour guide about the infamous 'Face on Mars', and they're sure to roll their eyes. It's nothing but an optical illusion, they'll say, a combination of geological fluke, chance lighting, and primitive photography. But even though this official version has been trotted out ever since the Face was first spotted, and no one sensible really doubts it any more, the face still exerts a strange fascination for first-time Martian visitors. Everyone, it seems, wants to see it at least once, as if to prove to themselves that it really isn't some vast monument of a lost Martian civilisation. It's fortunate then that the Face and its surrounding Cydonia region around it are well worth a visit on their own terms.

Viking's original image of the face. Some thought it looked human, some thought it looked alien. Other people thought it was some kind of human-lion hybrid. Right...

The Face was first spotted in 1976, lurking in photographs from the Viking 1 orbiter, one of the earliest robot probes to orbit Mars. NASA scientists were the first to notice it, and even pointed it out to the public as a joke.

However, they came to regret it when people started taking the Face seriously. What started out as innocent inquiry soon turned into a full-blown conspiracy theory, with NASA as the bad guys holding back evidence of intelligent

Looks familiar? Viewed from an angle, this face melts away quicker than Michael Jackson's nose in front of a roaring fire.

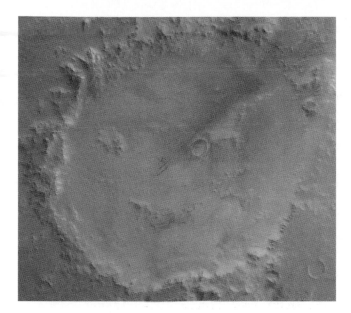

life on Mars. Despite a number of photographs from later spacecraft that clearly showed the Face as nothing of the sort, enthusiasts refused to give up, suggesting that it represented a human-lion hybrid, or even the face of the Sphinx at Giza, in order to explain why it didn't look properly human...

Even though the controversy about past Martian life has moved on to new and (in general) more realistic pastures, the Face is still an abiding image of Mars. Up close, it turns out to be just one of the many mesas dotted over the Cydonia region. They'd look familiar to anyone who's ever seen an old western movie filmed among the deserts and canyons of the southwestern USA. Mesas are outcrops of harder rock that stand alone after the rest of the landscape has eroded. On Mars, they might have formed from wind and water erosion over millions of years, could have been carved out by ancient glaciers, or may even have been pushed out of the ground by tectonic forces. The human tendency to make patterns means that many other 'structures' have been spotted in Cydonia, including a whole host of 'pyramids', and even the occasional 'lost city'.

Pyramids of Mars? Unfortunately not. Decades of exploration of the Cydonian mesas have yet to turn up any evidence that the Egyptians got here first.

Facing the storm

Martian dust storms have a fearsome reputation, largely the result of too many bad science fiction novels working their way into folk consciousness. Many tourists live in fear of being engulfed in the sands of Mars, never to be seen again, but the reality is rather different. With proper forward-planning, there's no reason to worry about getting caught up in one of the weeks-long global dust storms any more, and the worst a localized storm will do is slow down your itinerary for a couple of days, and (perhaps more of a problem) spoil some of your landscape photographs.

Most visitors agree that spectacular pictures of an oncoming storm front, and the traveller's tale for their gullible friends back home, more than make up for the inconvenience.

Assuming you'll want to catch at least one storm during your stay, what's the best way of second-guessing the

Martian weather? One good piece of advice is to hang around the northern-hemisphere plains. These dried-up ocean basins are dustbowls full of drifting powdery sands, and the Martian winds can really get themselves worked up after crossing the desert for a few hundred kilometres.

Dune fields

The same winds that periodically whip up dust storms act on a less violent scale too, herding wayward Martian dust across plains and crater floors until it catches around some slight obstacle and begins to build into a sand dune. The dune fields of Mars are frequently spectacular, with the dunes taking on bizarre shapes to match any displayed in Earth's deserts.

Martian deserts are also a surprisingly good source of water, unlike those on Earth. It turns out that the dunes themselves absorb water from the atmosphere. Ice helps to bind the sand grains together and strengthens them against erosion by the wind. It also allows them to hold together with steeper sides and even overhangs.

Frost comes and goes from the surface of a lot of Martian dunes, and of course ice can sometimes evaporate from inside. This can create local 'cracks' in the dune, and occasional fan-like depressions where parts of the surface slip downward.

DUST DEVILS

Much more common than the major sandstorms are the dust devils that skip across the flat plains of Mars. Although they look alarmingly like Earth tornadoes, the thin air renders them as underpowered as the dust storms, so don't worry too much if you see one heading your way. These Martian twisters also leave a trail behind them – look out for dark squiggles on the surface where they've vacuumed up the light surface sand and exposed darker material beneath.

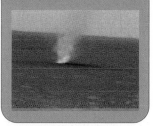

Light Martian winds whip fine sands from the peaks of Martian yardangs near the south pole.

Life on Mars

Is there, as late twentieth-century composer Sir David of Bowie memorably asked, life on Mars? He could hardly have suspected that, even in our modern age of space tourism and spandex jumpsuits, we would be little closer to an answer.

The quest for conclusive evidence of Martian life has been frustrating and fascinating. It began in earnest with the first spacecraft to land on the planet – the Viking probes of the 1970s. These carried an experiment to look for chemical reactions in the soil that might have indicated life. The results certainly showed that something unusual was going on, but, as usual, the scientists couldn't agree on what, and when the conspiracy theorists started to get involved (see also pp.92–93), it all got a bit messy.

In the 1990s, scientists realised that they didn't have to wait decades for a Mars probe smart enough to pick up a sample and return it to us – there are Martian rocks already lying around on Earth. A group working for NASA announced in 1996 that they had found unusual chemical traces in one of these Martian meteorites, and even showed photographs of what might be fossilized bacteria embedded in the rock. Opposing scientists soon found evidence to the contrary and a decade-long bunfight ensued, with both sides making claim and counterclaim. Like politicians and pub philosophers, there's nothing scientists like better than a good argument (and it's also good for research grants).

Successive landers and probes have added to the evidence that there may have been primitive bacteria in the ancient Martian soil, and that some might even survive today. For instance, there's a fair amount of methane in the atmosphere, and yet the Sun's radiation constantly breaks it up and blows its component atoms away into space. For it still to exist in such large amounts, something must be producing it today. The two candidates are active volcanoes and life, but the distribution of gas in the Martian atmosphere adds weight to the idea that it's life. Herds of flatulent Martian cows, anyone?

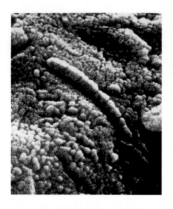

Looking a lot like worms, are these so-called Martian microfossils really ancient bacteria, or just mineral nodules?

LOOKING FOR LIFE

If you're into fossils or archaeology, you can always spend a few days as a volunteer with one of the various expeditions that scour the Martian surface for definitive proof of life. It'll likely involve a fair bit of digging for soil and ice samples, and that's hard work inside a spacesuit, but fame and fortune await anyone who strikes it lucky.

Know your Martians

The golden age of science fiction came up with countless memorable Martians. We're not suggesting that you're likely to run into these chaps during your tour of the surface, but it's just as well to be prepared...

Martian death ray

MARS ATTACKS

These big-brained gleeful killers from Tim Burton's 1996 movie were an affectionate spoof of countless earlier pulp sci-fi flicks.
Catchphrase: 'Ack ack ack ack ack!!!'
Survival tip: Carry an old gramophone and some easy listening music – Slim Whitman quite literally makes their brains explode.

Powerful pincers

Deadly heat ray

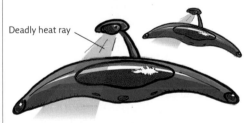

DOCTOR WHO

The Ice Warriors, invented for *Doctor Who* in 1967 and last seen in the recent revival fighting against his 33rd incarnation, were an ancient race that had hibernated due to Martian climate change.
Catchphrase: 'Ssssssso Dok-tor, we meet again!'
Survival tip: Carry a hot water bottle – they get weak from the heat.

THE WAR OF THE WORLDS

The ultimate Martians, dreamt up by H.G. Wells in 1898, and reinvented regularly ever since. Usually confined to their armoured tripod war machines.
Catchphrase: Not very talkative, but their war machines have a neat line in foghorns...
Survival tip: Keep a snotty tissue handy – they can't stand the common cold.

The polar caps

As we said earlier, you'll have to plan around the Martian seasons if you want to try winter sports here. Both poles have some underlying water ice, but they expand rapidly as carbon dioxide freezes out of the atmosphere in winter, and dwindle away equally swiftly as the ice evaporates in summer.

Both ice caps are impressive sights from orbit. Each sits on a mound that rises to several kilometres above the average Martian surface level, with high areas separated by winding, spiralling valleys, and scalloped cliffs around the edges. And each is embedded in a wider permafrost landscape: the high-latitude ground is a mix of soil and deep-frozen water ice, giving it the consistency of the

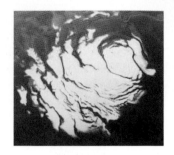

The beautiful swirling frosts of the Martian South Pole were a surprise to the scientists analysing the first pictures from the Viking orbiters.

Sunlight gleams off the permafrost around the north polar ice caps at sunrise in northern spring.

hardest rock. Where the permafrost has partially melted and then refrozen many times, bizarre geometric patterns can appear. If you didn't know better, you might think they were artificial (and plenty of people have suggested that they are!)

This is more territory for ice skating than skiing, but you'll need to catch the ice when it's at its thickest. That's during the long polar winter, when each pole gets several months of permanent darkness, and carbon dioxide from the atmosphere freezes straight onto the ground. During these periods, polar storms also sweep dust high into the air, where the tiny particles attract water from the atmosphere – just like hailstones forming on Earth. The crystallised water then sifts its way to the ground as dirty snow or hail, helping to build up the polar ice over a broad area of the surface.

Of course skating on an alien and sometimes dangerous landscape in the pitch black of a three-month night may not be the most appealing idea, so you're probably better aiming to get there for early spring, before the returning Sun has done too much damage to the ice.

When summer comes, the change is swift. Both carbon dioxide and that winter's fresh water ice begin to evaporate back into the atmosphere. At the north pole, the carbon dioxide disappears completely, but a little of each season's water frost survives, and the water that does melt away leaves behind dust from the atmosphere. In this way, the polar caps can grow a new layer of up to a millimetre each year.

With the snow crunching beneath your boots and the sunlight reflecting from the dazzling frost, the terraces around each pole are a beautiful sight. We still don't have much idea of why the terraces have formed in the shape that they have. One theory is that variations in the Martian climate affect the amount of dust mixed with the ice, and this means that some areas erode more easily than others. The overall pattern of spiral swirls, meanwhile, may show the direction of prevailing winds around each pole.

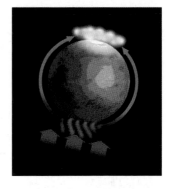

As sunlight warms the south pole of Mars during southern spring, carbon dioxide evaporates into the atmosphere. It's then carried to the north pole and deposited as fresh new frosts.

Further afield — Phobos and Deimos

Mars's pair of tiny moons don't have a lot going for them, to be honest, but since they're right on your doorstep in Martian orbit, it would be a shame not to visit, if only for the view. They also offer a quick sample of the asteroid belt without the need to actually go there.

Both moons are just a few kilometres across, irregularly shaped, and orbit very close to Mars – much closer than the Moon is to Earth. Deimos is the outer moon, just 15 km (9 miles) across, and orbiting at 23,500 km (14,580 miles) from Mars. Slip into orbit alongside it, and it's clearly a captured asteroid, though one with a few differences from other asteroids in the main belt (see 'Touring among the asteroids', pp.102-113 for some comparisons). For one thing

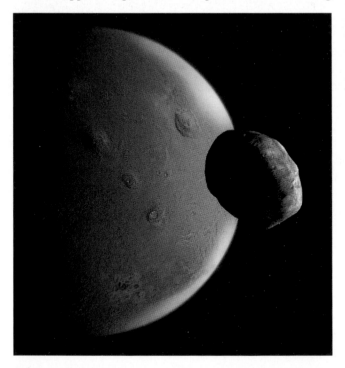

SWIFT'S PREDICTION

A literary mystery hangs around the Martian moons. It's probably nothing more than a coincidence, but you never know, and it bears repeating. When Lemuel Gulliver, hero of Jonathan Swift's 1726 novel *Gulliver's Travels*, visits the flying island of Laputa, he meets scientists who are up to various madcap or ill-advised schemes, including trying to extract sunlight from cucumbers. However, the Laputan astronomers tell Gulliver that Mars has two moons, and they even give their orbital periods with a surprising degree of accuracy. All this, of course, was 150 years before American astronomer Asaph Hall first spotted the two moons through his telescope in 1877. So did Swift have inside knowledge, or was it just a lucky guess?

Hovering in orbit alongside Phobos, it's easy to appreciate how the gravity from looming Mars could eventually destroy its inner moon.

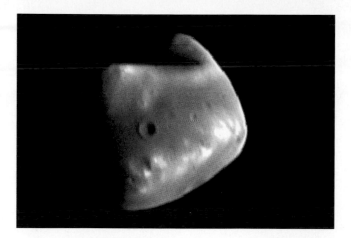

Small, lumpy, and a lot further from Mars than the more interesting Phobos, Deimos is low on most lists of Martian 'must-sees'.

it's quite a lot darker than the average asteroid, and for another, it's not as cratered. The same goes for Phobos, and that's led to the theory that maybe both moons originated much further out in the Solar System – they seem to be more like the Trojan asteroids that share Jupiter's orbit. Maybe that's where they started off, before something, perhaps a close encounter with the giant planet itself, threw them into a new orbit that eventually brought them close to Mars.

Anyway, Phobos, the larger moon, is 27 km (17 miles) long, and orbits even closer to Mars, skimming the planet at height of just 9,380 km (5,830 miles). That's so close that, despite its tiny size, Phobos can almost eclipse the Sun from the Martian surface. Its surface is dominated by a single giant crater, Stickney, some 10 km (6 miles) across. The impact that formed Stickney must have come close to breaking Phobos apart completely, and it might also be responsible for the moon's other weird surface features – long straight grooves or scars across its surface.

Enjoy Phobos while it lasts, because it's ultimately doomed. Mars and its inner moon have an intense but destructive relationship, as Phobos gradually spirals ever closer in. In about 40 million years' time, Phobos will plunge to a fiery end in the Martian atmosphere and the Red Planet will get a spectacular new crater.

THIRD MOON OF MARS?

Keep an eye on the long-range radar and be prepared for evasive action while in Martian orbit – it's just possible there's a tiny but homicidal third moon out there somewhere. The last photo taken by the Russian Phobos probe before it lost contact with Earth in 1989 showed a tell-tale streak across the picture. It could have been a camera glitch, but it might also have been a tiny, fast-moving moon zeroing in on the spacecraft.

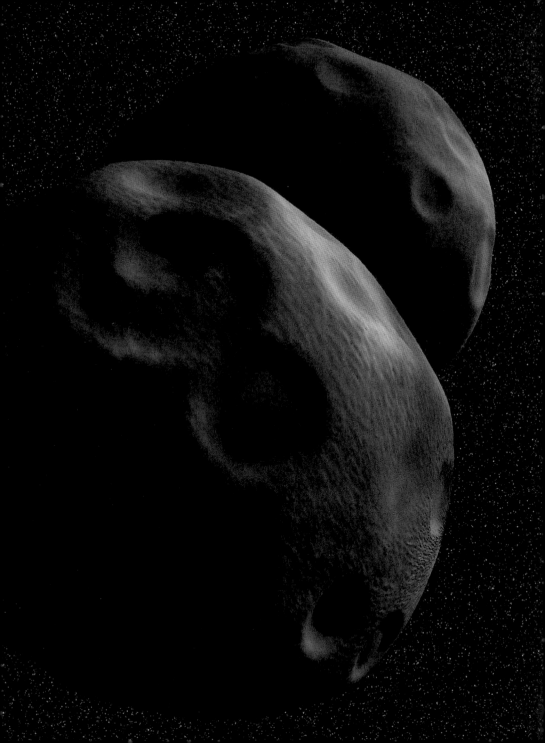

Touring among the asteroids

The **asteroids** are the Solar System's forgotten destination – literally millions of tiny worlds scattered across the inner Solar System, each with its own unique character. Naturally we can only focus on a few of them here, but be sure to take in Eros, an asteroid whose orbit crosses Earth's own, Ceres, the largest asteroid of all, and Vesta, a fragment of volcanic material from a mysterious larger world, long since vanished...

For those who want to holiday somewhere a little bit off the beaten track, the asteroids are an ideal destination. There are literally millions of places to choose from, so the chances are high that you'll be truly going where no being has gone before. Most of these rocky refuges are just oversized boulders, a few kilometres across at most, but a handful are much bigger – although nothing even approaches the size of our Moon.

The main asteroid belt sits neatly between the inner and outer planets, with its inner edge closely following the eccentric orbit of Mars.

A lot of people have the wrong idea about the asteroid belt, probably as a result of watching *The Empire Strikes Back* too many times in their youth. They imagine it's a near-impenetrable wall of floating rock separating the inner and outer Solar System, where kilometre-wide asteroids lurk around every corner intent on mowing you down like an unlucky squirrel on a four-lane motorway.

The sad truth is, the asteroid belt isn't nearly that much fun. To stick with the road comparison, it's more like a disused country track where the field mice can play among the grass for weeks without seeing so much as a passing combine harvester. It's not even a belt, really. While most asteroids hang around in roughly circular orbits between 254 and 598 million km (158 and 372 million miles) from the Sun, their variety of inclined orbits turns the 'main belt' into more of a fuzzy doughnut. They also have plenty of more adventurous brethren who wander off to see the bright lights of the inner Solar System or hang around with the big boys further out.

Getting there

With this huge spread, what should you aim for? It depends on how much time you have. For a short break on a limited budget, head for one of the Near Earth Asteroids whose orbits bring them conveniently close to our own. If you're taking a cruise and are out near Jupiter anyway, then a quick scooch along its orbit will let you take in a couple of the Trojans, small asteroids that hang around with the Solar System's school bully to look tough, but are careful to keep out of his range just in case they get hit by a flailing gravitational fist. But to get the most variety in a relatively short time, and perhaps as an ideal excursion from a Mars-based holiday, head for the main belt.

ASTEROIDS ON EARTH

Asteroids are among the few celestial objects that you can study up close without ever leaving Earth, because they rather obligingly come to us. While most of the meteors that burn up in Earth's skies are mere specks of dust left behind by comets, the meteorites that make it through to the ground are nearly all fragments of asteroids. Geologists have lumped meteorites into several different families, such as the 'carbonaceous chondrites', 'irons', and 'stony irons'. Comparing and matching the properties of the various meteorite types with the more distant asteroids gives them a head start in understanding the nature of the asteroids themselves.

CELEB SPOTTING

Asteroid names may have started out in the same high-minded, classical vein as those of other planets and moons, but it soon became clear that there were so many of the things that they could use up every pantheon of gods in every major religion, and still barely scratch the surface. As a result, the naming of asteroids was made into a free-for-all: you find it, you get to name it. These days, you can have fun zipping round the asteroid belt looking for asteroid 9007 James Bond, 17059 Elvis, 8748 Beatles, and even 18610 Arthur Dent!

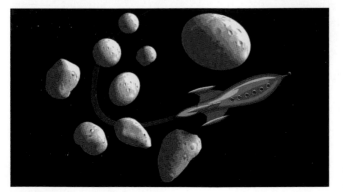

Unless you fly too close to a recently broken-up rubble pile, you won't need to worry about dodging asteroids.

Ceres

The largest asteroid by a long way is Ceres, a ball of rock 960 km (596 miles) across with a roughly circular orbit about 414 million km (257 million miles) from the Sun. Fittingly, it's asteroid number 1. It contains about one third of all the material in the asteroid belt and is massive enough for gravity to pull it into a ball shape, but apart from that, it seems pretty much unchanged since the early days of the Solar System.

As you approach Ceres, you'll notice how dark the surface is. It's also heavily cratered: clearly Ceres has taken quite a battering throughout its history. Getting into orbit is a delicate business – Ceres' gravitational field is pretty pathetic, at just one sixth of the Moon's, or 1/36 of Earth's.

That same weak gravity means you can have a lot of fun bounding around on the surface. A gentle hop can send you tens or even hundreds of metres into the air, taking several

Heavily cratered Ceres is the only asteroid large enough to pull itself into a sphere.

minutes to come back down. On the ground, you'll see that the horizon is noticeably curved and much closer than you might expect. It's quite an experience to take a really powerful leap and watch this pockmarked world fall away beneath you – this must be what it feels like to actually *be* a rocket.

Of course, the flipside is that you have to be careful moving about. You won't weigh enough to do yourself much harm, but some people find it quite unnerving if every step sends them almost into orbit. We recommend that you take grappling hooks and anchor yourself firmly to the ground. That way you can pull yourself in if you accidentally launch yourself too far from the surface.

And what about that dark surface? It turns out that Ceres is rich in 'carbonaceous' minerals, a mix of carbon-based chemical compounds similar to the coating of comets. There are even signs that there was once ice here – some of the minerals contain water, locked away in crystalline form.

Vesta

STAR ATTRACTION

The enormous crater at Vesta's south pole exposes the asteroids interior, and reveals how it's separated into different layers.

Asteroid 4, Vesta, is quite a different proposition from Ceres, and is certainly well worth a visit. At 560 km (348 miles) wide, it's a little over half the size of its bigger sister, though still one of the largest asteroids in the belt. It's also unusually bright for an asteroid, bouncing back far more sunlight than anything else in this part of space. Combined with an orbit slightly closer to the Sun (at an average distance of 353 million km (219 million miles), this makes Vesta the only asteroid visible to the naked eye from Earth (just... if you have good eyesight... on a dark night... and you know where to look).

You should be able to spot a couple of Vesta's most remarkable features from some way out. As it rotates every 5.3 hours, its overall brightness varies quite a bit, suggesting that the surface is mottled with a mixture of light and dark patches. Through binoculars, you should also be able to see that Vesta isn't a perfect sphere. Now, Vesta is one of the few asteroids whose gravity should be up to the task of pulling it into a proper spherical shape, so it's not just a case of another low-gravity, irregularly shaped asteroid. Indeed, the northern hemisphere seems perfectly spherical, it's just the southern half of the planet that's badly misshapen. A massive collision at some point in Vesta's past has created a huge crater at the south pole, complete with a towering central peak.

Hover over the crater, or even land if you have the time, and you'll get an insight into Vesta's big secret, the reason it's so unusually bright. The impact has gouged a hole in the asteroid's crust and surprisingly Vesta turns out to have a crust (and a mantle to expose underneath). That's both impressive and puzzling for a world this small. It means that Vesta must have melted all the way through at some point in its past, allowing heavier elements to settle into its centre, just like in a full-sized planet such as Earth. But this never happened on Ceres or any of the other large asteroids.

That's not all, though. The crust that covers the rest of the planet turns out to be made of bright and relatively shiny basaltic rock (no dirty carbonaceous compounds here). And that rock can only have covered the surface in one way we know of – through volcanic eruptions.

So what the heck is going on with Vesta? We're still fairly baffled by it, to be honest – internal layering (scientists call it 'differentiation') and volcanic activity should both be out of the question for a world this small. One explanation would be that Vesta is a leftover fragment of some long-vanished world, perhaps knocked off its parent planetoid in an enormous collision that melted all its material. This would have morphed it into its new layered structure and eventually, as its crust solidified while the interior remained molten, could have briefly created volcanoes on the surface. The alternative is that Vesta ended up with a different composition from all the other asteroids by some other means, and this allowed it to evolve into a mini-planet while all the other asteroids stayed stuck in their primeval state.

ATTACK OF THE VESTOIDS!

Astronomers got to analyse bits of Vesta long before their spaceprobes reached it. It turns out that there's a whole group of smaller asteroids, plus quite a few meteorite fragments found on Earth, that are dead ringers for the surface of Vesta. Geologists like to call them 'basaltic achondrites', but that's just scientific jargon for rocks made of basalt that don't contain chondrites (the chunks of carbon-based material found in a lot of meteorites and asteroids). Personally, we think 'vestoids' is much catchier.

Breaking up is not so hard to do...

A slight nudge was all it took to shatter an already-fragile rubble pile asteroid. Each fragment is now on an independent orbit around the Sun, but gravity may eventually pull most of them back together again.

If there's one thing we've learned from our explorations of the asteroid belt, it's that a lot of asteroids go to pieces faster than a starlet at an awards ceremony. The main belt is full of the evidence for this. Match orbits with a typical asteroid and you'll likely find yourself sharing the space with a variety of waifs and strays that have been blasted or knocked loose from its surface at some point in its past.

Our final recommended stop in the main belt is 253 Mathilde, a pretty average asteroid, very roughly spherical, and about 66 km (44 miles) across. It takes a skilled pilot to get into orbit around this chunk of rock, because the gravity

Asteroid mining

In the next few decades, asteroid mining could go from being the subject of endless speculation and the occasional rash investment, to a booming space-based business. Already a variety of corporations are jealously eyeing up some of the smaller and more convenient asteroids, poring over pie charts that show their estimated mineral content and value. So if you land anywhere, you'd better make pretty sure you don't get tangled up in an ownership dispute!

According to the most enthusiastic prospectors, once you overcome the basic problems of mining in low gravity and getting material back to Earth, it should work out a lot easier than digging tunnels and blasting quarries on Earth.

The main reason for this is that most asteroids are pretty much unchanged since their formation. On Earth, exposure to air, water, and endless recycling forces locks many valuable elements away in complex minerals. To get them out again, you'll often have to use a tedious and expensive chemical process named after some nineteenth-century chemist or other. On an asteroid, however, elements usually stay in their natural form – or at worst combined in fairly simple compounds. Once asteroids can be broken up, sifted, and ferried back to Earth orbit, the enthusiasts assure us (with just a hint of fanaticism) that it will be the beginning of a new, space-based industrial revolution.

So if in doubt, check your charts to see if anyone's staked a claim before landing. Asteroids don't yet have automatic defence systems, but you could find yourself on the receiving end of a nasty writ if you're not careful!

---

is so weak. There's another clue to this in the sharp edges on its craters. There's very little ejecta around to soften the edges, because it all flew off into space when the craters were made.

The funny thing is, Mathilde's surface looks perfectly solid, but it's strictly for looking, not touching. Looks are deceptive, and Mathilde's gravity reveals that its about as dense as water, on average. Since the bits we can see are rock, the whole asteroid must be riddled with empty space to explain it having such low mass. Mathilde's typical of a whole class of asteroids that are little more than orbiting rubble piles, clinging together so tenuously that a gentle kick might set them on a slow road to eventual disintegration.

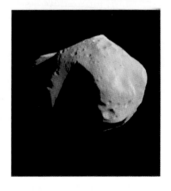

Mathilde is thought to be roughly 40% empty space. If you break it, you can put it back together again!

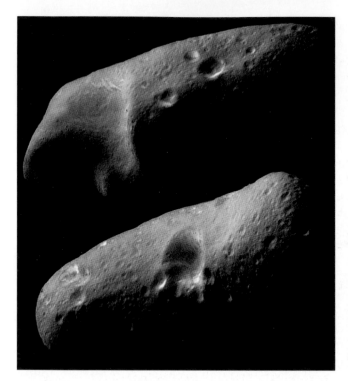

Eros looks very different depending on your angle of approach. In the top view, the 'saddle' feature is obvious, while the 'cat's paw' crater is at the centre of the lower image.

IMPACT!

The first thing everyone wants to know about Near Earth Asteroids is how likely they are to hit Earth. In Eros's case, the statisticians reckon there's a one in ten chance of it colliding with either Earth or Mars in the next million years. Those might seem like pretty scary odds, but we're dealing with astronomical timescales here. Try to remember that the whole of human history, plus prehistory back to the dawn of consciousness, fits neatly into the last hundred thousand years. Asteroid orbits are notoriously difficult to predict for more than a few dozen passes round the Sun, but these days, it would be relatively easy to strap a rocket engine to any threatening asteroid and gently nudge it out of the way. It's the ones we might not see coming, most likely comets plunging towards Earth from the Oort Cloud (see p. 202), that we really should be worrying about.

Further afield – Eros

As we've already mentioned, there are plenty of wayward asteroids out there, roaming the space between the major planets and generally looking to start trouble. Probably the most famous, and also one of the easiest to reach from this part of the Solar System, is Eros, asteroid number 433.

Don't be fooled by the name, though. If you're looking for a romantic getaway, this place is a bit of a dump, and your significant other is unlikely to be impressed at the thought of spending a weekend in orbit around a barren piece of space junk

That didn't stop an early NASA space probe, the Near Earth Asteroid Rendezvous, from spending a whole year here, though, Of course, NEAR was a robotic probe, and back then, robots didn't have much in the way of imaginations, but it did arrive at Eros on Valentine's Day of the year 2000, and made a final romantic gesture, 'kissing' the surface at the end of its mission, exactly one year later. As a result, Eros is one of the best studied asteroids around. Just 31 km (19 miles) long, it's shaped like a moderately deformed potato, with an obvious indentation called the 'Saddle' halfway along its back.

Dropping into orbit around Eros, you might be surprised at the strength of its gravitational field. Even though it's about half the size of Mathilde, its gravity is stronger. That's because Eros seems to be a fairly solid lump of silicate rock, similar in density to parts of Earth's crust. There certainly aren't any huge hidden voids like the ones within Mathilde.

Eros's gravity is certainly strong enough to pull back some of the ejecta from the occasional meteor impacts on its surface. As a result, much of the surface is strewn with boulders, which have rolled in strange directions thanks to the asteroid's irregular shape. Go down to the surface and take a walk to see what we mean: it's bizarre to stand on one side of the Saddle and follow the opposite curving wall up to almost straight above your head, and it's even weirder that you can then stroll effortlessly up that vertical wall, to find the spot where you previously stood is now overhead.

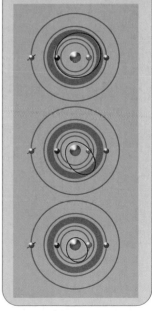

This photo was sent back by the NEAR probe during its final descent to the surface of Eros in 2001.

The ultimate package

The Solar System's biggest planet offers its biggest holiday experience. Visit **Jupiter** and you get not only the ever-changing gas giant itself, but also the huge system of moons that surround it, a veritable mini-Solar System of fascinating worlds. Take a flight into Jupiter's atmosphere, see the spectacular sulphur volcanoes on Io, go skating on Ganymede and fly over the Solar System's biggest nature reserve, the ice-covered ocean of Europa.

Astronomers of ancient times gave Jupiter a head-start in the competition for tourism when they named it after the ruler of the ancient Roman gods (also known as Jove, hence the adjective Jovian). Its huge family of moons, including four remarkable giant satellites, gives Jupiter probably the biggest 'bang per buck' of any Solar System destination. The only problem you may have is getting there – in Earthbound terms, a worthwhile trip to Jupiter is more of a round-the-world cruise than a weekend city break.

Getting there

Any spacecraft capable of getting you to Mars should also be fit for the longer trip beyond the asteroid belt, but you may want to invest in something more powerful in order to speed up the journey time. A word of advice here, that will stand for all the outer planets – when talking to your dealer or hire firm, make it clear where you're heading. As well as decent main engines, you'll also need some good manoeuvring thrusters, since a proper tour of any of the major moon systems will involve a fair amount of planetary swingball as you switch between one orbit and another.

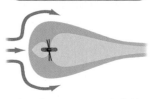

Jupiter's immense magnetic field parts the solar wind, and creates a 'magnetotail' that stretches to the orbit of Saturn.

Inside Jupiter

Jupiter and the other giant planets have very different interiors from the rocky worlds of the inner Solar System. The surface cloud layers occupy just the outer hundred kilometres or so of the planet. Beneath this, it's almost pure hydrogen and helium all the way down. As pressure increases, the gases get compressed into liquid form. At greater depth, molecules of hydrogen get broken apart into atoms, creating an ocean of liquid metallic hydrogen. At the centre of it all is a core about the size of the Earth.

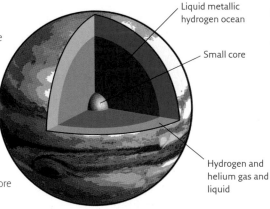

Liquid metallic hydrogen ocean

Small core

Hydrogen and helium gas and liquid

During final approach, the Galilean moons are frequently silhouetted against Jupiter itself.

One of the pleasures of the journey is watching as Jupiter gets bigger and bigger in the cabin window from week to week. It's so huge that it almost makes a perceptible disk from Earth, and by the time you're swinging into an initial orbit, still some 16 million kilometres (10 million miles) from your destination, Jupiter will be a beautiful sight, the size of the Full Moon and ideal for viewing with binoculars or a telescope. As you get further in, you'll start to lose this planet-wide perspective as the outer edges of the disk become foreshortened and hidden.

From a high orbit, it should also be clear that there's something odd about Jupiter's shape – it's squashed like a semi-deflated stripey beachball. Actually, squashed is the wrong word, since it's really the equator that's bulging out, not the poles sagging in. The reason for this is what most people still wrongly call centrifugal force (it's not actually that simple, but this isn't a physics textbook!) Because Jupiter is so massive and spins on its axis in just 10 hours, it's a little like an ice skater turning a pirouette with a bucket of water in each hand – the speed of the spin naturally throws Jupiter's midriff outward. The same thing happens to the Earth, but Jupiter's larger diameter, faster spin, and gassy constituency make it far more obvious here.

JUPITER DATA

Good points:
A huge variety of worlds in one location

Bad points:
Dangerous radiation belts
No landings on Europa

Day length:
9 hours 56 minutes

Year length:
11.86 Earth years

Cloudtop gravity:
2.6 g

Surface temperature:
-110 °C
-160 °F

Communications time:
35 minutes or more

Close approach

The Great Red Spot is a huge outward bulge in Jupiter's cloudscape, drawing red chemicals up from lower in the atmosphere.

A long spiral in towards the planet will give you ample time to enjoy the spectacular view, but don't just concentrate on Jupiter itself. The four major 'Galilean' moons, all roughly the size of our own Moon, are an entrancing sight as they waltz around the planet in orbits varying from 42 hours (for innermost Io) to 16.7 days (for outer Callisto). Astronomers have been watching their dance for centuries, and they played a key role in the revolution that kicked off modern astronomy (see 'Galileo and his moons', p.121), but from Jovian orbit, the view is something else. Assuming you're approaching across the ecliptic plane (and the fuel economics don't really allow much else), you'll see how the moons frequently eclipse one another, pass into Jupiter's shadow and behind its disk, and later move across its face, casting their own shadows onto the cloudtops. Look out for the mysterious 'post-eclipse brightening' of Io. It's another of those now-you-see-it, now-you-don't phenomena that give astronomers night sweats, an occasional and inexplicable boost to the brightness of Io as it emerges from Jupiter's shadow back into the sunlight.

Another elusive effect may stretch your eyesight, but is at least scientifically understood. As you pass round to the

MYSTERY SPOT

Time your visit right, and the Great Red Spot will live up to its reputation, but at other times, the Spot is neither Great nor Red, and can be distinctly underwhelming. Earth-based astronomers have been watching its comings and goings since at least 1830, and possibly 1655. Before we knew that Jupiter had no solid surface, some people thought it might be a floating island in the currents of a global ocean. We still don't fully understand how it works today.

night side of the planet on your approach spiral, see whether you can locate a thin crescent Io, and look for a yellow glow around it. This is the sulphur torus, a doughnut-shaped ring of trapped particles blasted out of Io's gravitational grasp by its famous volcanoes, but trapped by Jupiter's magnetic field. In ideal conditions, it's sometimes possible to trace the torus all the way around the planet.

By the time you're 100,000 km (60,000 miles) out, Jupiter will really dominate your view, its huge looming bulk occupying almost half of the sky. From here, you should be able to see the hypnotic and ever-changing detail of the cloud systems. For ease of reference, astronomers often split the majority of Jupiter's cloud features into two types – light-coloured zones and darker, red or blue belts. The terms date back to a time when we knew far less about Jupiter, and they give precisely the wrong idea about the planet – namely that the zones are where we see the underlying lighter clouds, and the darker belts are wrapped around on top of them. In fact, it's the lighter clouds that are higher up, and the darker regions that offer a glimpse into the soup of the lower atmosphere.

Jupiter's weather works in pretty much the same way as Earth's. Clouds form in regions of low pressure, and are driven away from regions of high pressure. One difference is that Jupiter is nothing but clouds, all the way down, so the high-pressure regions just part the lighter, high-altitude clouds. The other major difference is that Jupiter's rapid rotation acts like your washing machine's spin cycle, flattening the different regions out and wrapping them round the 'drum' of the planet. It's called the coriolis effect, and the same thing happens to a lesser extent on Earth, where it creates the system of prevailing winds in each hemisphere. Because different layers of the atmosphere travel at different speeds, the belts and zones effectively move in opposite directions to each other.

RADIATION HAZARD!

While in orbit, try to steer clear of Jupiter's huge array of radiation belts. The planet has an intense magnetic field, many times stronger than Earth's so it accelerates the particles trapped within it to faster, deadlier speeds. If you're planning a reasonable tour of the Jovian system, you're going to be orbiting in this natural microwave oven, so make sure your spacecraft and spacesuits are both fitted with high-grade shielding.

Different colours in Jupiter's cloud layers correspond to different heights. Blue are the lowest, then brown and cream. Red clouds such as the Great Red Spot form at the top of the atmosphere.

Slipping between Jupiter's cloud layers is an unforgettable experience, but be careful to keep clear from the massive thunderheads!

Bobbing around like marshmallows above the lightest cloud layers are huge oval storms – enormous low pressure areas that are usually capped by very bright, white, rotating clouds. These often start to form at the boundaries between belts and zones, where the movement of the atmosphere in opposite directions creates complex currents. Most of the time, the cloud is just whipped into decorative curls called 'festoons' – the kind of thing that wouldn't look out of place on a well-iced wedding cake. But if an air mass wraps around itself enough, it can become a full-fledged anticyclone, a self-regenerating cell of circulating low-pressure air, spinning anticlockwise while trapped between the two atmospheric bands. It's a bit like a diabolo, where rotation started just by moving the string in two directions eventually sets the thing spinning fast enough to fly free.

ONCOMING STORM

So-called 'white spot' storms come and go frequently from Jupiter's surface, but occasionally they can combine into something bigger. When storms merge, they gain strength as well as size. The largest storms create such low pressure that they rise for kilometres above the surrounding atmosphere, and can create updrafts of chemical soup from deep inside the Jovian atmosphere. The most famous example of this is of course the Great Red Spot – here the pressure is so low that high-altitude red cloud, found nowhere else on Jupiter, can often condense.

Into the clouds

A descent into Jupiter's cloudy upper atmosphere is an unforgettable journey, even though it carries some inevitable risks (but then, if you've got this far, you've probably got a healthy sense of optimism anyway). Check in advance that your spacecraft is atmosphere-rated, because the outer envelope of gas starts well above the visible cloud tops. You'll also want to check you have good engines, since you'll be aerobraking, losing speed as well as altitude all the way down. Unless you want to get be trapped in Jupiter's gravity forever, you'll need to have your exit planned.

The recommended regions for a trouble-free descent vary almost as much as the weather on Jupiter. The boundaries between belts and zones may look pretty, but the countercurrents created in the atmosphere make them, if not lethal, then certainly hazardous for all but the most experienced Jovian sailors. One popular option, on the other hand, is a flight along the cloud canyons that open up on either side of the Great Red Spot. These are comparatively quiet zones, stretching for thousands of kilometres.

Alternatively, use a thermal imager to study the planet's heat output, and head for any obvious hot spots in Jupiter's clouds. These areas, where heat escapes from the planet's interior, are usually marked by calm weather and clearings in the upper cloud layers.

There's only so far you can go into Jupiter's atmosphere if you want to come back. Within a couple of thousand kilometres of the cloudtops, the pressure builds up to the point where it'll crush your ship like a paper bag. If the hull starts to creak, it's time to get out quick. Remember that it's almost impossible for a spacecraft hull to be equally good at resisting outward pressure in the vacuum of space, and inward pressures that rival those experienced by deep-diving submersibles.

GALILEO AND HIS MOONS

Italian scientist Galileo Galilei is probably the pivotal figure of Renaissance science. He stuck his nose into everything and, of course, made some of the first telescopes. He first got wind of the new Dutch invention in 1609 and soon built his own, allowing him to discover star clouds in the Milky Way, the phases of Venus, and mountains on the Moon. Most importantly, though, he spotted the four 'Galilean' moons of Jupiter. This convinced him that maybe the theory published by an obscure Polish priest called Copernicus in 1543 was right, and not everything moved around the Earth after all. To be honest, Copernicus's idea of a Sun-centred Universe was catching on in northern Europe already, but the Catholic Church was still convinced that the Universe revolved around the Earth, and Galileo was one of the first people with the nerve to start preaching this Copernican nonsense on their doorstep. Despite being an old mate of the Pope, he ended up on the wrong end of an inquisiton, and spent his last years under house arrest.

Over the night side

As you pull away from Jupiter's cloud canyons, take a look at a different aspect of the giant planet with a flight over the night side. The sight of its bulk blocking out the distant Sun is unforgettable and ominous, but once you've got used to the difference, you'll realise that night over Jupiter isn't as dark as you'd think. The clouds are constantly flickering as mighty bolts of lightning jump between them. The effect is a bit like a nightclub fight illuminated by strobe lighting, and the sight of Jupiter's clouds swirling as they flicker on and off is almost as likely to induce nausea, so don't stare for too long.

The polar regions offer an altogether more sophisticated light show, in coruscating colours. Rings around the poles mark the area where the Jovian magnetic field and atmosphere meet, and solar wind particles trapped and accelerated by the magnetic field plunge gleefully into the upper layers of gas. There's also another, closer source for particles – some of the material belched out by the volcanic moon Io meets its end here. The results are aurorae (northern and southern lights) far more intense than any seen on Earth. Detune your shipboard radio, and you'll be able to hear the static hiss that comes and goes with the intensity of the storms.

The other highlight of the night side is only visible from further out. Look beyond the darkened limb and you should see a thin line stretching out into darkness. This is Jupiter's rather paltry ring system, a flattened disk of material that starts around 20,000 km (12,500 miles) above the cloudtops and stretches out as far as the orbit of Thebe, 245,000 km (150,000 miles) from Jupiter. Most of the particles that make up the ring are microscopic, so small that you can plough through them on the way to and from the planet itself, and probably won't notice a thing. The rings only become obvious when you can see them backlit by the Sun while Jupiter's disk is dark, so it's little wonder that they weren't discovered from Earth. In fact your view of them is almost identical to the original discovery photograph, taken by Voyager 1 in 1977.

HOT JUPITER

Thermal imaging cameras trained on Jupiter's night side will reveal another secret. The planet pumps out more heat than a fat guy on a treadmill. In fact, it emits a lot more energy than it receives from the Sun in the first place. All the giant planets do this, with the exception of Uranus (but then, Uranus is weird in other ways too).

Where does the energy come from? The best guess is that it's due to friction at work on a massive scale. The giant planets have probably never stopped contracting from their initial formation, though now the contraction happens mostly deep below their clouds. As heavier elements gradually sift towards the centre, they bump past their neighbours in outer layers, transferring some of their energy and helping to keep the entire planet warm.

A view over Jupiter's night side takes in the backlit rings, polar aurorae, and the crescent of the planet itself.

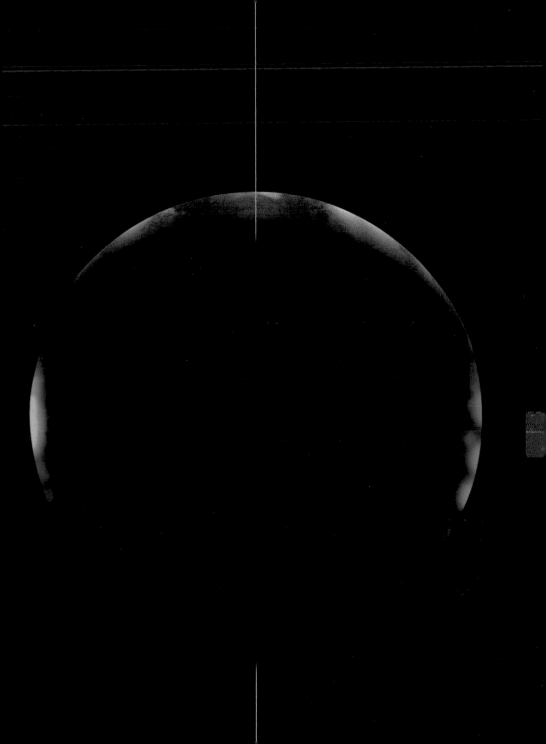

Io: among the volcanoes

Io is the Jovian system's star turn, a tortured volcano world and the closest of the major satellites to the planet itself. Many travellers still treasure their sulphur-stained space-suits as souvenirs of their visit, but there are a few who bear the scars of their encounter with this violent and unpredictable world, and a handful who never made it back.

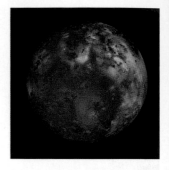

Io's volcanoes stand out like livid bruises, while its sulphur frosts create lighter terrain.

The reason for all the activity on Io is that the moon is periodically squeezed this way and that in the vice-like grip of Jupiter's gravity. The hotter regions of the interior, the core and mantle, are much more malleable than the surface, and have become the interplanetary equivalent of one of those rubber balls that are supposed to relieve the stress of overworked executives. Just like them, the interior gets hotter as it's constantly pushed and pulled around, and this has allowed Io to keep its molten interior when it would otherwise have cooled and frozen solid long ago. The effect is called tidal heating, and as we'll see elsewhere, it makes the moons of the outer Solar System a lot more interesting than you'd expect.

As Io moves round its orbit, its changing distance from Jupiter means it is alternately squeezed and relaxed, heating its interior.

Io's widely acknowledged as the most volcanic world in the Solar System, with several dozen active volcanoes at any given time, but don't expect to see the kind of volcanism you'd find at Mount Vesuvius or on Hawaii. Io's innards contain a lot of

Io hangs in front of Jupiter's clouds in this image from the Saturn-bound Cassini spaceprobe, which flew past Jupiter in 2001.

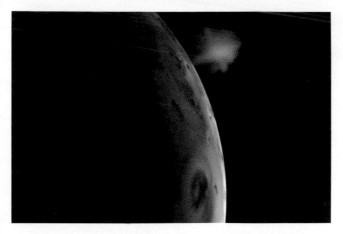

Sulphur plumes are easiest to see when Io itself is a crescent and most of the sunlight is coming from behind the moon.

sulphur-based chemicals, which are much lighter than the basaltic, iron-rich lavas of Earth. Sulphur also has some unique chemical properties that allow it to take a variety of forms, ranging from powdery yellow stuff to sticky brown gloop and black tar. The surface of Io is a mix of all these forms and more (technically they're called allotropes). Add in a variety of basic compounds such as sulphur dioxide, and it's the recipe for a very colourful world, though one that'll make you pine for the blues and greens of Earth.

From orbit, Io looks like a mouldy pizza, but on the surface it's more like the aftermath of a fire in a candle factory. Sulphur erupts from the surface in a variety of ways, ranging from bubbling volcanic calderas to powdery fountains. Over time, thick deposits build up around the vents, gunging them up until they seal over, splutter and die. Blocked-up former gushers lend the landscape an eerie, organic quality.

Land near one of the volcanic plumes around sunrise or sunset, and you should get a spectacular view as the volcanic spray is illuminated in the last rays of sunlight. The ground around the plumes is generally covered with layers of yellow-white, crunchy sulphur 'frost', and if you hang around for more than a few minutes, you'll probably get a coating yourself. It's formed as powdery sulphur spat out in the plume drifts slowly back to the surface in Io's weak gravity.

GEYSERS AT WORK

Io's sulphur plumes, like a lot of volcanoes in the outer Solar System, work on the geyser principle. Hot magma pushing its way up from inside the planet meets a relatively cool seam of sulphur dioxide trapped underground. The sulphur dioxide is quickly heated to its boiling point, but because it's trapped below ground, it can't boil. So the 'superheated' liquid runs along under the ground until it finds a weak spot. As soon as it burns through to the surface, it boils explosively into the near-vacuum of Io's atmosphere, with such force that it arcs hundreds of kilometres over the surface, and some of it never comes down.

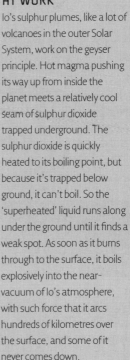

Pele is one of Io's most spectacular volcanic vents, and also one that's historically important. It's named after the Hawaiian volcano goddess, and was the first active volcano to be found on Io (or indeed, anywhere beyond Earth). Voyager 1's mission scientists were surprised when their pioneering spaceprobe turned its cameras back as it retreated from Io, revealing huge plumes of material rising above the limb of the planet. It was still erupting with just as much power when the Galileo probe, Jupiter's first orbiter, got there nearly twenty years later, and seems to be a powerful geyser with a near-inexhaustible supply of sulphur dioxide 'fuel'.

Lava lakes

The most spectacular features on Io are without a doubt the volcanic calderas, huge lakes of sulphurous lava that periodically open up in Io's surface. It's generally safe to take a low flight over them, but if you want to go on foot, you'll need to be careful. Often the ground in these regions is little more than a thin skin of crusty sulphur that has built up around the edge of the caldera. Even in Io's weak gravity (just under one fifth of Earth's), you'll need to be wary of where you put your weight.

There are at least 200 calderas more than 20 kilometres (13 miles) across dotted over the surface of Io.

Io TECTONICS

It might come as a shock, but not every feature on Io is volcanic. Because its interior is so hot, there are also some low-level tectonics going on. There's no sign of the crust actually splitting into plates, but there are regions where neighbouring areas of the landscape have apparently been pushed in opposite directions (presumably by convection cells churning away in the mantle below). These 'crumple zones' have pushed up some big mountain ranges. Our personal recommendation is Tohil Mons, a spectacular peak 5,400 m (17,700 ft) above the average surface, at the centre of an upland region 300 km (185 miles) wide.

Jupiter's massive bulk rises over the sulphurous Ionian landscape.

They seem to come in two varieties. Most contain lakes of molten sulphur at relatively low temperatures (though they'll easily burn through your spacesuit in seconds, so don't get cocky). A few are scaldingly hot and contain molten silicate rock – not so dissimilar to some Earth-based volcanoes. Molten lava escapes from these calderas in flows across the surface in pretty much the same way as on Earth.

With all its spraying, oozing, erupting and generalized leaking, Io resurfaces itself at a phenomenal rate. The current best guess is that it throws up enough material each year to cover the entire landscape to a depth of around 5 mm (1/5 inch). Even though the eruptions are localised, there's still enough material to change the global geography quite quickly. If you've got a particular place in mind to visit, our advice is to do it quickly – put it off for a few years, and it could be unrecognisable, and even untraceable. Also, make sure that your maps are up to date!

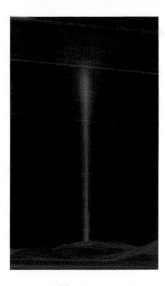

Be wary of getting too close to a sulphur plume. Although they cool rapidly and are harmless to fly through, on the surface it's a different story.

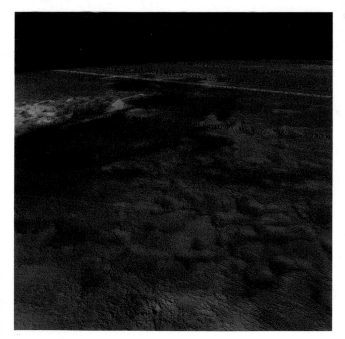

Io's volcanic calderas reshape the landscape around them in a way best appreciated from low orbit.

Europa

Jupiter's second major moon draws almost as many visitors as Io, but icy Europa is strictly for looking and not touching. It's been a conservation area, off limits to tourists and kept as pure as possible, since the very beginning of the Space Age. So if you're reading this and don't fancy ten years retraining as an astrobiologist and filling out mountains of paperwork, you'd probably better resign yourself to enjoying the view from low orbit.

Even from a distance, Europa stands out from the other moons. While they are mostly browns and greys, Europa is white with a slightly pinkish hue. As you get closer, the pink features start to resolve themselves into criss-cross scars running over the surface. Closer still, and you'll start to notice something else. Although Europa has a lot of surface features, it's as flat as a particularly flat pancake. There are no mountains here, no deep valleys, and even the impact craters seem to have suffered a bout of nihilism, their rims slumping back to the surface as if to say 'what's the point?'

One popular metaphor for Europa is to say that it's as flat as a billiard ball. However, a better way to picture it is to imagine the Earth with no hills rising above a couple of

A distant image of Europa reveals a multitude of cracks over the surface that nevertheless have little effect on the moon's overall smoothness.

Inside Europa

Europa's icy crust may extend for as much as 10 km (6 miles) beneath the surface before giving way to the global ocean. The sea itself could be up to 100 km (60 miles) deep, heated from below by the stretching and compression of the planet's rocky innards. Most of the heat is generated in the semi-molten crust, and there could be undersea volcanic vents at the top of the mantle. At the centre lies a solidified rocky core.

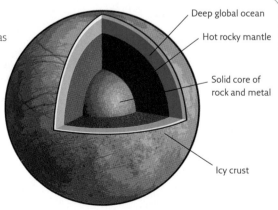

Deep global ocean

Hot rocky mantle

Solid core of rock and metal

Icy crust

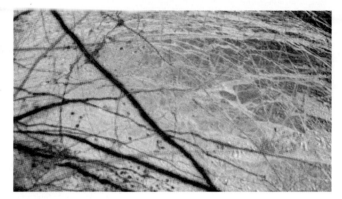

An enhanced-colour photo from the Galileo space probe shows the differing ages of the surface. Dark blue is old, pure water ice. Red shows where fresh, chemically stained water has welled up through the cracks.

hundred metres, no deep seas or mid-ocean trenches... nothing much, actually.

Not exactly a recipe for an interesting world, you'd think? Except this very smoothness is a sign of what's really going on at Europa. The big clue comes from the craters and the way they've flattened themselves out despite the fairly weak gravity. This indicates that Europa's surface isn't all that it seems. Also, there are so few craters compared to other worlds in this part of the Solar System that clearly something's been at work resurfacing Europa.

Dive as close as you can to Europa's surface (the thin oxygen atmosphere won't trouble most spacecraft), and if you've seen Earth's old polar caps, you'll probably get a sense of déjà vu. From a few tens of kilometres up, Europa's surface looks just like pack ice. Not the deep-frozen solid ice you'll see on other moons, but loose-packed, jostling fragments that look as if they're ready to move at any moment.

They're not, of course – try looking for the sea on which they float and you'll be disappointed. If there was any exposed water at the surface of Europa, the near-vacuum conditions would cause it to boil away into space, in spite of the sub-zero temperatures out in Jovian orbit.

That doesn't mean that the water isn't there, though – it's just locked away. Evidence for a global sea beneath Europa's ice is now pretty conclusive, although geologists are still tussling over the precise amount of water and the depth of ice

SHIFTING ICES

Although Europa's surface is not as loose as drifting pack ice, it's not frozen completely solid either. Currents in the water below help to pull blocks of ice in different directions, and the helping sledgehammer of Jupiter's gravity attacks any weaknesses, opening up occasional cracks in the surface. As water wells up from beneath, it boils away into the atmosphere before freezing solid and plugging the gap. But while Europa's skin heals itself remarkably rapidly, it leaves the planet with scars. The fresh ice is stained with chemicals from the sea below, giving it a distinctly pinkish hue.

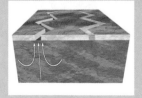

beneath which it's buried. The basic idea is that, as with Io, Europa's core gets a pummelling from Jovian gravity that leaves it battered, bruised, and hot enough to drive volcanic eruptions, raising the temperature of the blanketing ice sheet enough to melt it completely.

Interesting though all this geology is, it wouldn't normally be enough to declare an entire world off limits. The reason for that is far more exciting: scientists want to keep Europa in a pristine condition because it's one of the most likely places in the Solar System to shelter life.

Why would a deep-frozen, almost airless snowball so far from the Sun be a good place to look for the neighbours? The answer's in the water. For centuries, scientists used to think that all life on Earth depended on sunlight for its ultimate energy source. But in the late twentieth century, they began to discover how wrong they'd been. Some of the Earth's nastiest nooks and crannies turned out to be teeming with so-called extremophile organisms. These drop-outs from the great surface struggle took one look at the sunlight and decided that it was overrated, opting instead for alternative lifestyles eked out around deep-sea vents, in hot volcanic rocks, and in acidic groundwater (and who are we to make judgements about someone's lifestyle choice just because they eat rock or metabolise sulphur for a living?)

On Earth, the volcanic-vent communities of the deep ocean floors are the most spectacular and successful of these

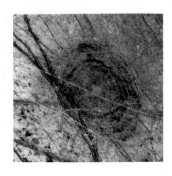

The ripples of this ancient impact crater have slumped as Europa's ice has shifted, leaving just a fingerprint-like ghost in the ice.

Jupiter rises over the serene whites and pinks of Europa's icy crust.

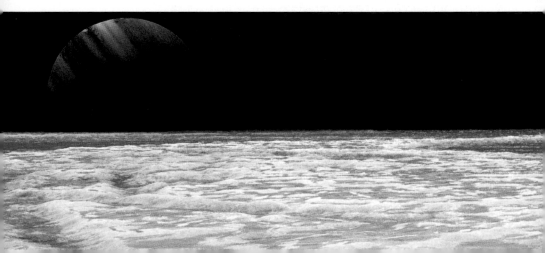

hippie lifeforms. Way back in evolutionary history, a lucky worm found that colonies of sulphur-eating extremophile bacteria in its stomach could produce all the nutrients and energy it needed. This was the ancestor of the huge tube-worms that anchor themselves in vast gardens around many vents today, providing food and shelter for a whole host of other drop-outs that have adapted to a new food chain.

So if it happened on Earth, why not on Europa? The conditions around volcanic vents here should be just as good as those on Earth, so the big question is now whether life needs sunlight to get started on its evolutionary marathon. Even if it does, who's to say that Europa might not have been seeded with a deep-frozen starter kit for life from the occasional comet impact? This 'panspermia' idea is certainly cherished by a number of astrobiologists.

It may be that we'll never know for certain. Almost as soon as they discovered Europa's secret ocean, the old space agencies started working out how they might one day send a probe through the ice and say 'hi' to any intelligent molluscs down there. So far, all these plans have come to nothing, and Europa's waters have remained hermetically sealed. Maybe one day, we'll get lucky and a crack will open up into which we can drop a free-swimming probe. For the moment, though, we'll just have to be content to admire this pristine, mysterious world from a distance.

KEEP EUROPA TIDY

One of the big problems with Europa is keeping it uncontaminated. Earth bacteria are tough little so-and-sos, and for some of them, a half-billion kilometre trip across space is no more hassle than a trip down the shops for a packet of fags. If we ever do find hints of life on the Europa, scientists don't want to spend decades squabbling over whether it came from a bit of dog dirt trodden into the lab by a technician several decades ago. And of course, if there is life here, we don't want to subject it to nasty bugs from Earth. Just look at what European diseases did to the natives of central America.

The rising hemisphere of the Solar System's largest moon clearly reveals the patchwork of different terrains on its surface.

Ganymede

Although its smaller, Jupiter-ward siblings tend to hog the limelight, no visit to the Jovian system would be complete without a tour of Ganymede, the largest moon in the Solar System, and a fascinating world in its own right.

At first glance, it's easy to understand why Io and Europa get the star treatment while big brother Ganymede is left in the shadows. It's a mostly grey and white world, with heavy cratering in some areas and none of the obvious signs of activity seen on Io and Europa. Get closer in, though, and it's a different story. Parts of the planet start to look like a frozen winter lake, and that's not too far from the truth. Other bits also look like an attempt at decorative plastering, almost as if someone's dragged a comb through the surface before it set properly, creating long strips of parallel ridges and valleys.

From orbit, you'll notice that Ganymede's got quite a strong magnetic field, suggesting that it might have a molten core at its centre, like Earth. No big deal, you might think: Io and Europa certainly have a core, mantle and crust, and Ganymede's bigger than them, so should be hotter.

Here's the but: Ganymede doesn't receive a fraction of the tidal workout that Jupiter gives to Io and Europa.

ANOTHER OCEAN?

Does Ganymede also have an ocean beneath its crust? It seems pretty likely, but the evidence isn't quite so conclusive as it is for Europa and Callisto (see p.136). To be honest, Ganymede seems a far more likely candidate for a salty layer of water than Callisto. The giant moon's natural magnetic field clouds the evidence a bit, but it still seems very likely that there's an ocean just below the surface, perhaps 160 km (100 miles) down and several kilometres deep. Anyone who takes a big enough drill with them could probably book their place in the scientific textbooks now...

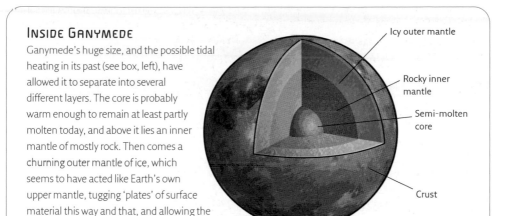

Inside Ganymede

Ganymede's huge size, and the possible tidal heating in its past (see box, left), have allowed it to separate into several different layers. The core is probably warm enough to remain at least partly molten today, and above it lies an inner mantle of mostly rock. Then comes a churning outer mantle of ice, which seems to have acted like Earth's own upper mantle, tugging 'plates' of surface material this way and that, and allowing the grooved terrain to form between them.

Icy outer mantle

Rocky inner mantle

Semi-molten core

Crust

Okay, it's still big, and so should be better at retaining heat, but the closest real comparison, Callisto (see p.136), isn't much smaller, and seems to have stayed a frozen ball of dead rock ever since it formed.

Anyway, before we blow the plot completely, let's see what the surface has to offer. Most people head for the ridges first, since they look like the more interesting bits, but to get a proper feel for them, we'd say you'd better start among the darker patches.

Memphis Facula is an interesting spot, a 350 km (220 mile) crater in the dark terrain with a smooth icy floor known as a 'palimpsest'. We'll see these here and on Callisto – the name comes from an old word for paper that has been wiped clean and reused. What probably happened here is that a large meteorite smashed into Ganymede and made a big hole straight through the crust into the slushy ice below. As ice rushed in to seal the gap, it froze solid, creating a bright and shiny new surface.

This seems to have happened a lot on Ganymede, and for the largest impacts of all, there was a second effect. Fly over the Galileo Regio area and you'll notice deep troughs in the dark surface. The patterns can be hard to trace with all the

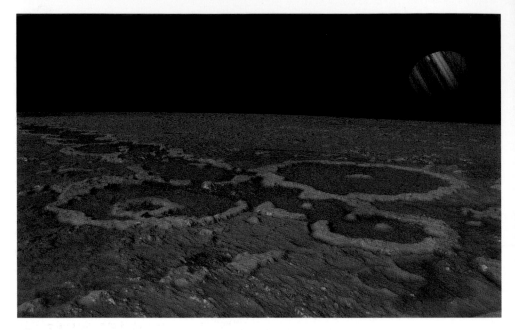

later cratering on top of them, and the brighter groovy terrain separating them, but the furrows seem to be arranged in concentric ring patterns, a bit like crater walls.

When subsurface ice rushed in to fill the major impact craters, it seems that it dragged bits of the crust along for the ride. The crust didn't travel far before slamming on the brakes, but it would have separated enough to create deep faults, allowing some bits of the surface to slip below others.

Ganymede's location makes it a prime target for meteorite and comet impacts. As Jupiter's gravity pulls innocent passers-by to their doom, the Galilean moons are right in the line of fire. See, for example, the spectacular Enki Catena crater chain, more than 160 km (100 miles) long, and formed as the broken fragments of a comet smashed into the moon.

Enki Catena records the impact of a necklace-like chain of comet fragments that intecepted Ganymede while heading for Jupiter.

Uruk Sulcus

Sulcus is the Latin name for what we prefer to call 'groovy terrain' the brighter bits of Ganymede that look as if they've been combed over. The name just means a groove or a trench anyway. Uruk Sulcus is among the most impressive of these areas – fly across it at low altitude, and you'll be able to appreciate the dazzling range of grooves running across each other in half a dozen different directions.

The interesting thing about the *sulci* is that they're a mix of old and new terrain, suggesting that a region of the surface has been broken up, pulled apart, and had new icy material added to patch the gap. It's quite a difference from Europa, for instance, where a simple fracture opens up, new material fills the hole, and it's often possible to see where the older feature lies on both sides. On Ganymede, it's more common to find one half of a feature intact, and the other side lost in the jumble of a *sulcus*.

A closer look, or an expedition to the ground, will soon show that not all *sulci* are the same. In some cases, chunks of the crust have simply collapsed downwards as cracks opened up on either side of them. This type of terrain is called a graben, and is more likely to have formed where Ganymede's crust slowly pulled apart. Elsewhere, though, the separate ridges are tilted blocks of crust, resembling dominos that have toppled over and rest on top of each other. These probably collapsed at an angle where the crust was separating at a much faster rate.

EXPLAINING GANYMEDE

So why are Ganymede and Callisto so different? It turns out that tidal heating is responsible after all. A billion years ago, Io and Europa can united their gravitational forces to bump Ganymede off its cosy circular orbit, into a more elongated one that put it through a lot more stress and tidal heating.

Until then, Ganymede was just another craterball, but heating melted the interior, allowing it to separate into layers. Churning in the icy mantle started to pull the crust around, creating faults that eventually split apart in parallel lines. Ice welled up through the caps, and the *sulci* were born.

Sulci run through Ganymede's more ancient craters, showing how they formed through stretching of the moon's crust.

Callisto

If Io and Europa are the John and Paul of Galilean moons, and Ganymede is George, then Callisto must be Ringo. It's often unjustly written off as the boring one, mostly because it's a comparatively dark world, and much more heavily cratered than its inner neighbours. But Callisto (like Ringo) has hidden depths...

From half a million kilometres or less, though, Callisto wouldn't look out of place hanging from the ceiling of an old-time disco – a real glitterball of a moon. It seems that the dark surface coating goes only a little way down, and that any half-decent impact can puncture this surface layer, revealing fresh bright ice from just below the surface, and blasting it out as rays of ejecta.

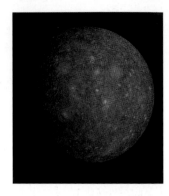

From a distance, Callisto has little to recommend it apart from seas of craters – its surface has changed little since ancient times.

Distance from Jupiter means Callisto doesn't get the full-on tidal pummelling dished out to the inner moons, and doesn't even have the kind of relationship to the other moons that could result in a temporary burst of tidal heating, as happened to Ganymede. As a result, Callisto was left frigid, its interior an unsorted jumble of rock and ice. With little or no geological activity to help resurface the landscape, the moon has preserved almost every splash from several billion years of steady crater rain. Some people reckon it's the most heavily cratered body in the entire Solar System: its large size and position on the firing line to Jupiter mean it's seen more impacts than an army of crash test dummies.

The most impressive crater of all is the Valhalla Basin, an enormous bullseye of frosty white amid the darkness. The walls of the basin form bright rings across 2,600 km (1,600 miles) of the moon's surface, while the central area is one of Callisto's several palimpsests. These are the equivalent of lunar seas areas where a large impact allowed slushy ice to well up from inside the planet and cover the crater floor. They're just about the only bit of Callisto where older craters have been wiped away.

THE WATERS OF CALLISTO

Callisto's ocean gives itself away through the way it behaves in Jupiter's enormous magnetic field. The interaction between salty water and magnetism creates a weak 'induced' magnetic field. It's easy to tell apart from the kind of magnetic field produced by iron in the core of a planet or moon, and anyway we know that such a core is out of the question for Callisto.

Almost everything on Callisto comes back to cratering, sooner or later. Take the bizarre 'spires' to the south of the Asgard impact basin, for instance. These towering pinnacles of bright ice surrounded by dusty debris look like nothing so much as termite mounds, but even they can be explained through cratering. They're chunks of dirty ice ejecta from the impact that formed Asgard itself, buried deep in the landscape millions or even billions of years ago. Heat from the Sun has gradually burnt off much of the ice, eroding them into the current pinnacles, and allowing the rock and dust to slump off the sides and lie around their bases.

Callisto's pinaccle terrain is a good candidate for the title of 'most bizarre landscape in the Solar System'.

Callisto's biggest secret, though, is locked away below the surface. Bizarre though it seems, this frigid world somehow maintains an ocean of liquid salty water beneath its surface. The first hints of this came from the Galileo Jupiter probe in the late twentieth century, and it's likely that the ocean is about 200 kilometres (125 miles) down, and between 10 and 100 kilometres (6 and 60 miles) deep.

Amalthea

Study any chart of the Jovian system and you could easily overlook Amalthea. But in our opinion it's a must-see destination, if not for the lump of rock itself, then for the view, which must be one of the most spectacular in the Solar System.

Amalthea is 262 km (163 miles) long and 150 km (94 miles) wide. On first approach, it looks a lot like a large asteroid, peppered with craters of various sizes, including the huge 90 km (56 mile) Pan. But a proper look at the surface reveals that it's anything but – the rock here suggests that it was once part of a larger world. With so many comets and asteroids pulled to their doom by nearby Jupiter, it's little wonder that at least one of its moons found itself in the wrong place at the wrong time. Much of the moon was able to pull itself together again, but the reformed Amalthea is still full of holes.

Today it's fascinating to stand on this wreck of a world and just enjoy the view. From here, Jupiter's cloudtops are just 110,000 km (68,000 miles) away. The planet occupies fully one third of your field of view, and you can watch the swirl of its cloud patterns and the dancing shadows of the Galilean moons at relative leisure, provided you're well protected and dosed up against radiation.

Amalthea's got one other mystery. As you crunch around on the surface, you might notice that the crust has a distinctly red colour – in fact, it's one of the reddest objects in the Solar System. Even more strangely, some steep slopes such as crater walls, have a distinctly greenish hue. One possible explanation is that Amalthea sweeps up coloured sulphur falling towards Jupiter from Io, but then why wouldn't its surface be yellow and orange? It's possible that Amalthea's permanent exposure to Jupiter's radiation belts changes the colour of material from Io that drifts onto it, giving the small moon its predominantly red appearance.

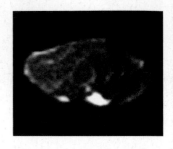

From a distance, Amalthea looks like a pretty average, irregularly shaped inner moon, of a type you can find by the dozen around the giant planets.

The view from Amalthea is the most spectacular of any Jovian moon, with Jupiter occupying the entire horizon in one direction.

Lord of the rings

Some people dismiss **Saturn** as a less-interesting Jupiter, but it has much more to offer than that suggests. Descend through its outer haze and you'll find a world as colourful and active as Jupiter. Play zero-gravity hopscotch between house-sized boulders in the fabulous ring system. Visit Mimas, the Solar System's real-life 'Death Star'. Climb the razorblade crags of Hyperion and make a landing on Titan, the strange frozen world where methane behaves like water does on Earth.

The Solar System's ringed wonder is probably the most exotic destination still within (relatively) easy reach of Earth. Beyond Saturn's orbit, vacation turns into exploration, and only the hardiest independent travellers are found. However, Saturn itself still attracts large amounts of tourist traffic, drawn by the delights of its spectacular rings, and by a system of moons just as exotic as those around Jupiter.

Even if you do have to approach when Saturn's rings are edge-on, the shadows cast on the planet make a wonderful photo opportunity.

Getting there

But even with a high-performance spacecraft, the return trip is not for the faint-hearted. Saturn is roughly twice the distance of Jupiter, and getting there quickly is best accomplished by a gravitational slingshot around Jupiter itself. Talk to a travel consultant about possible flightpaths – although your rendezvous with Jupiter will be a brief one, if you plan it well you should still be able to have a good look at the Jovian system, and you may even be able to zip past one of the major moons. Also, look out for other flyby opportunities to enliven the trip. A small asteroid or comet can keep most people amused for a few days, and believe us, by the time you reach Saturn you'll be sick of Travel Scrabble!

Saturn is second only to Jupiter in size, and it orbits the Sun once every 29.4 years. If you plan to take a scheduled tourist flight, keep in mind that demand and prices tend to dip every 15 years. If that seems odd, bear in mind that Saturn is tilted on its axis, rather like Earth. Twice in each orbit its equator (and the ring plane above it) line up with the plane of the Earth's orbit (across which you'll be approaching). Because the rings are so thin, they disappear almost completely during these 'ring plane crossings' It doesn't make any difference once you get there, but it can certainly make Saturn look a lot less interesting during the last couple of months of the journey. Photography nuts, in particular, regard these crossing periods as poison. They (and most other people) prefer to approach the planet during Saturnian summer or winter, where one pole or the other tilts towards the Sun, and the rings are displayed 'wide open' to Earth and approaching spacecraft. It may be cheaper to travel in the off season, but just as on Earth, the result can be some disappointingly dreary holiday snaps...

Because the rings are so fantastic, Saturn itself is often written off as Jupiter's rather plain cousin. This isn't entirely fair – in fact the phrase 'pale and interesting' could have been invented for this sepia-tinted world, as you'll soon find out.

SATURN DATA

Good points:
The rings, of course. A huge variety of moons

Bad points:
Quite a way to travel

Day length:
10 hours 39 minutes

Year length:
29.37 Earth years

Cloudtop gravity:
1.1 g

Surface temperature:
-140 °C
-220 °F

Communications time:
71 minutes or more.

THE RING THING

That wily old Italian Galileo Galilei (see p.121) was probably the first to spot that something wasn't quite right with the shape of Saturn, though the weakness of his telescopes stopped him figuring out exactly what was going on. In 1610, he announced that Saturn was accompanied by two large moons, while other astronomers thought it had 'handles'. Both theories came unstuck in 1612, when Saturn went edge-on to Earth and the rings disappeared. It wasn't until 1655 that Christiaan Huygens correctly suggested that the planet was surrounded by a thin ring.

The view from orbit

Flying in alongside the D ring (see p.147) in a low-inclination orbit. Note the pale shadows cast across Saturn's opposite hemisphere.

Saturn's cloudscape is a creamy-white version of Jupiter's. As you get closer to the planet, you'll start to see that it is almost as active as Jupiter itself, it's just that Saturn's weather systems tend to involve off-white storms running between magnolia cloud belts, with the occasional sepia swirl for company. It's almost as if Saturn was a TV set and someone's turned the colour right down...

In fact, that's not far from the real explanation. Saturn's atmosphere is quite similar in chemical terms to Jupiter's, and the apparent difference comes from a high-altitude layer of white misty cloud that drains away all the colour, while not quite hiding the weather features themselves.

These high-altitude clouds are formed by ammonia crystals in the cold upper atmosphere. Jupiter could in theory have formed a similar outer fog layer, but because it is closer to the Sun and has a more powerful internal heat source, its

RING AVOIDANCE

Inclined orbits around Saturn are only safe if you stick close to the planet or very far out, in areas where the ring particles are too small to do much damage. Get your orbit wrong so you intercept the main rings, and you could spend the last few seconds of your life dodging house-sized ice boulders.

outer layers are too warm. Saturn, by the way, also has one of these internal generators, probably powered in the same way as Jupiter's by the slow sifting of heavier chemicals down into the core.

If you fancy a dip in Saturn's atmosphere, you'll probably want to avoid the weather conditions on the equator itself. Here, a steady rain of microparticles sifts down out of the rings, burning up in the atmosphere. Try to get into an orbit that crosses the equator at a fairly shallow angle, and you can keep this storm of microscopic hail to a minimum while still witnessing an impressive display of uncountable faint meteors as you fly in alongside the plane. As always in space travel, it's better to be safe than to be sorry, especially when being sorry can be a brief, fatal experience.

One problem with shallow-inclination orbits that keep you close to the equator is that they don't really give you the full Saturnian experience. Using a highly elliptical orbit, however, you should be able to change your orientation economically (at the far end of the orbit you'll be travelling much more slowly, so your engine burns will have a lot more effect). Alternatively, you can always plot a slingshot encounter with one of the inner moons to change your course.

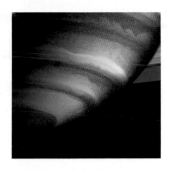

Saturn's atmosphere has frequent electrical storms. In this false-colour picture, the red glow close to the top is a vast thunderstorm.

INSIDE SATURN

Saturn's internal structure is pretty similar to Jupiter's, with an upper cloud layer, and a clear atmosphere of hydrogen and helium gas slowly transforming into liquid. Where the pressure grows high enough to break molecules into atoms, an ocean of liquid metallic hydrogen is created, surrounding a core of rock and metal at the centre. The major difference from Jupiter is that the outer layers are comparatively thicker, since the planet's overall gravity is much less and the upper atmosphere can expand outwards.

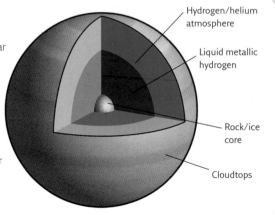

Hydrogen/helium atmosphere

Liquid metallic hydrogen

Rock/ice core

Cloudtops

Flying in from the poles is in many ways a more impressive experience, since from this angle the rings will be wide open to you, spread across the sky. Depending on the time of Saturnian year, either the north or south side of the ring plane will be in direct sunlight, turning it into a brilliant, if monochrome, rainbow arc. Towards one horizon or the other, the rings may suddenly disappear in mid-air, hidden from sunlight by the shadow of Saturn itself cast across them.

As you fly in towards the equator, the rings tower higher and narrower in the sky, until, almost too quickly to notice, they've narrowed to invisibility and begun to widen again. The 'unlit' side of the rings is in fact only slightly fainter than the sunward side: the ring particles are so tightly packed that they reflect a considerable amount of light onto the dark sides of their neighbours. As the rings widen and sink back towards the horizon, watch for their passage in front of the Sun. Here, you'll be passing through the shadow of the rings cast back across the planet's surface. Mllions of particles of various sizes cause the Sun's light to flicker rapidly, and some travellers report that the entire experience makes them nauseous.

Ringworld

STAR ATTRACTION

In recent years, Saturn's rings have become a playground for adventurous tourists, a spectacular fluke of nature that has created a zero-gravity theme park for the well-insured. Here, you can hang impossibly between house-sized blocks of ice

ROCKHOPPING SAFETY

Experienced ringwalkers recommend that anyone planning to spacewalk in Saturn's rings completes a basic safety course first. Even then, they suggest that ringwalkers remain tethered at all times – if not directly to their spacecraft, then at least to a large ring fragment within easy reach of safety. It's easy to forget yourself while hopping from rock to rock, and there have been several incidents where tourists used up the fuel reserves for their manoeuvring units, leaving them stranded beyond the reach of their spacecraft, and leading to expensive and dangerous rescue missions. For this reason, it's also a good idea to deposit a fuel reserve on your 'base' rock before beginning to explore in earnest.

thousands of kilometres below Saturn's looming bulk, play orbital hopscotch using the smaller chunks as stepping stones, or simply admire the spectacular views from in and around the ring plane.

The rings vary wildly in size and composition, and all are worth visiting to capture their unique moods. Closest to the planet lies the D ring, a near-invisible plane of microscopic ice and dust particles that stretches all the way into Saturn's atmosphere. A combination of atmospheric drag and the effect of Saturn's magnetic field ensures the D ring is in constant flux. As particles fall out of the ring towards the planet, they are replaced by new ones from further out.

Next comes the C ring, often called the Crepe Ring. Crowded with particles a few centimetres in diameter, the C ring is almost diaphanous, allowing you to look through it, at the spectacular bulk of Saturn below.

But the main rings, A and B, are where the excitement really starts. This is dangerous territory, where house-sized boulders of ice tailgate each other like rush-hour drivers around the orbital freeway, frequently jostling for position and occasionally colliding and spinning out of control. From a distance, the A and B rings are opaque, and they cast deep shadows on the surface of Saturn. Up close, you'll see they consist of countless narrow individual rings nested within each other, looking very like the grooves on an old vinyl record from the early days of sound recording. Occasionally, small clearings appear in the rings – spokes

A collision or breakup in orbit creates a cloud of debris in a variety of random orbits. Collisions between particles orbiting in different directions soon flattens the debris cloud out into a series of concentric, circular rings.

The major rings of Saturn. From inside to outside, the D ring, C or Crepe Ring, B ring, Cassini Division, and the A ring (including the narrow Encke Divison).

radiating out from Saturn itself, or wave-like ripples, both caused by the gravitational effects of Saturn's inner moons.

However, it's not until you're nearly on top of the rings that you can see the individual particles. They're remarkably confined within a single plane that sometimes just a hundred metres or so thick. Collisions or close encounters with some of the nearby moons sometimes send ring particles spinning out of the plane, but any unruly runaways are swiftly brought back into line by the beating they receive on each subsequent passage through the ring plane. The basic rules of the road are pretty much the same as on Earth – stay in lane, or expect to get hit!

Piloting in the ring plane can be a hairy experience, so we generally recommend you stay just above or below it – it's surprisingly easy to hover just a few tens of metres away and yet be perfectly safe. If you must go in, however, try to stick close to one of the moonlets, the larger chunks of rock and ice around 100 metres across. These giants are the highway patrol cars of the ring system: their gravity makes everyone else keep a respectful distance, and as a result they create a clearing in front and behind them.

Spacewalking into the ring plane is an exhilarating (and nerve-wracking) experience. It feels as though the oncoming boulders could swat you aside or leave you stuck to their tyres like so much roadkill, but you've got one key advantage over the hedgehog crossing the road: provided you match orbits correctly, you'll be travelling at the same speed as the orbital rock garden. From your point of view, everything will be eerily still. Most of the ring fragments barely even rotate, locked in Saturn's tidal grip so they keep one side towards the planet. If you do come across any large spinning boulders, it's probably a sign that they've recently been involved in a collision.

With no gravity to hold you back, it's an amazing experience to jump from one boulder to another across the void of empty space. Make sure you're manoeuvring equipment is fully charged, however. Ideally, you should also carry grappling equipment to get a grip on the surface.

MYRIAD MOONS

Even after Huygens had correctly described Saturn's rings (see p.143), it took another two centuries to figure out what they actually were. Many astronomers were convinced that they must be solid objects, but the forces acting at different distances from Saturn would soon tear such a huge structure apart. Others thought they might be fluid of some sort. Eventually, however, James Clerk Maxwell, a Scottish physicist chiefly known for his boring but important laws of thermodynamics (the bane of students to this day...), arrived on the scene. Maxwell correctly figured out that the rings had to be composed of countless small objects, each in its own orbit around Saturn, and confined to a single plane and a fixed radius by the gravity of the planet and its moons.

Moonlets without number stretch around Saturn in this view from just above the B Ring.

Mimas

Any seasoned traveller's photo album of Saturn is bound to include a few choice snapshots of Mimas. Not only does this moon offer a superb platform for viewing Saturn from close up, but it's also a bizarre world in its own right. This relatively tiny satellite, just 418 km (256 miles) across, is dominated by the huge Herschel Crater, a dent in the surface so large that it stretches from the equator almost to the north pole. This gives it a startling resemblance to the Death Star from the classic *Star Wars* movies.

After receiving such a large impact, it's a little surprising that Mimas is still around to tell the tale. It's also a good demonstration of what the moons of giant planets have to put up with, orbiting stoically like ducks in a cosmic shooting gallery while their parent planet pulls in any passing object that feels like taking a pot-shot at them.

Herschel is a great place for a low fly-by. With no atmosphere to drag on your ship, you can zip past just a few kilometres above the surface. In fact, the crater is probably best appreciated from even higher up – get too low

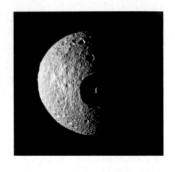

Herschel Crater hovers ominously on the terminator between day and night in an image straight out of *Star Wars*.

As the closest major moon to the rings, Mimas is a dream for photography nuts. Here, the tiny satellite hangs in front of ring shadows on Saturn's northern hemisphere, while the rings themselves cross beneath it.

down, and Mimas's short horizon will make it impossible to see the whole thing at a glance.

Saturn and its edge-on rings loom above the walls of the huge Herschel Crater on Mimas.

For the full Mimas experience, we recommend touching down with a low-gravity buggy close to the anti-Saturnian point on the far side of Mimas, the region of the moon that faces precisely away from the centre of Saturn. There's not much to write home about (the surface is blanketed by craters of all sizes, and that's about all) but the fun is to be had from the journey, not the destination. Trek towards the Saturn-facing side along the equator, and you'll be treated to one of the Solar System's most amazing sights. You won't have to go far before you'll notice a stiletto-like blade of light rising up from the horizon, the edge-on plane of the ring system. If you can wait till sunset, it's worth trying to spot the glow of the much fainter and broader outer rings.

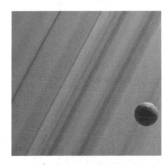

Mimas hangs against the blue-tinted winter hemisphere of Saturn, crisscrossed by ring shadows.

Gradually, the ring rises higher in the sky, until it almost bisects it. Now Saturn itself begins to appear, a bloated yellow-white orb heaving itself over the horizon. It seems to hang impossibly close, and an approach round the equator makes it seem all the more bizarre since the entire planet seems tipped on its side, with the cloud layers falling vertically down towards the horizon like rippling curtains. It's truly one of the highlights of the Saturnian system, and all the better when seen across the huge basin of Herschel Crater itself.

Enceladus

The first thing you'll notice about Enceladus (pronounced 'EnSELadus') is how bright it is. Twenty per cent bigger than Mimas, with a diameter of 512 km (318 miles), Enceladus is a brilliant white moon, reflecting 70% of the sunlight that reaches its surface.

Now the Solar System is a mucky place. Meteorite strikes tend to dirty things up pretty quickly, and comets are no better – yes they're snowballs, but they're snowballs made from the kind of brownish mulch you find left in the middle of the road two days after the snowfall. So anything as bright and white as Enceladus is downright suspicious. There's something going on here, and a sneak peak through binoculars will confirm it. Enceladus has hardly any craters, so it's clear that something has been at work cleaning up the surface very recently.

An enhanced-colour image of icy Enceladus clearly shows the blue 'tiger stripes' that are the source of its eruptions.

Swinging over the night side, keep a look out in the skies above the dark edge of the planet. If you're lucky, you'll see the blackness turned milky white, almost as if this tiny moon somehow had an atmosphere. Flying through these rising plumes of material is nothing to worry about, and in fact if your spacecraft has one of those whizzy new particle analysis rigs on the hull, this is somewhere you can give it a test run. Turns out the plumes are mostly crystallised water ice, with a bit of carbon dioxide and methane thrown in for good measure. In everyday terms, that's snow.

In Enceladus' weak gravity, the plumes shoot their icy contents a long way. Some material escapes the moon's grasp altogether. Spread out around its original orbit, it becomes Saturn's extremely faint and sparse E Ring. Most of it, though, falls back onto Enceladus itself as snow, turning the landscape into a permanent winter wonderland.

To see these plumes close up, we recommend landing in the southern hemisphere. An image enhancement camera should highlight the area you're looking for – bluish 'tiger stripes', almost invisible to the naked eye, run across much

NAMING MOONS

Saturn's moons don't have quite the same classical ring to them as the ones around Jupiter, but that's probably just because we know some stories better than others. While Jupiter's moons are all named after his various mythological lovers and other hangers-on, Saturn's are nearly all named after the Titans, a legendary generation of Greek gods who ruled, squabbled and fought before the classical gods led by Jupiter rose to power. Confusingly, the Titans were actually all children of Uranus, not Saturn, in the myth.

of the southern hemisphere. With a thermal imager, they'll stand out like a sore thumb, since the ice in these stripes is about 20 °C (36 °F) warmer than the bulk of the planet, at a comparatively balmy -180 °C (-290 °F).

Land close to an active stripe around dawn or sunset, and you'll be able to appreciate the sight of an ice plume catching the sunlight high in the sky, without suffering too much from the brightness of the landscape. If Enceladus was much closer to the Sun, we'd be recommending ski goggles against snowblindness, but this far out, even such a reflective world is comparatively dim.

Gravity here is about 1% of Earth's, so provided you've exercised during the journey, you should be able to bounce around like a hyperactive toddler. Playing hopscotch through the plumes is reminiscent of streaking through a garden lawn sprinkler, the usual silence of the spacesuit suddenly disrupted by the drum of water droplets bouncing off its outer layers. The plumes themselves won't do you any harm, but be careful where you land, and try not to hit the same spot twice. The ground is already cracked and weakened, and we don't want you disappearing down an ice crevasse!

This image, captured by the Cassini spaceprobe in 2006, was the first to show one of Enceladus's powerful geysers in action.

Saturn rises over the icy blue landscape of Enceladus's southern hemisphere. Note the erupting ice geyser on the right.

Three icy moons

Sandwiched between the geysers of Enceladus and the mysteries of Titan, the icy worlds Tethys, Dione, and Rhea are all too easily overlooked. But they do have some impressive sights to offer, and they can tell us quite a lot about the way Saturn's moons formed and developed.

For one thing, there's an obvious pattern in sizes going on among Saturn's moons. Mimas and Enceladus, which we've already looked at, are roughly the same size, and Tethys and Dione are both about twice the size of either of them, with diameters of 1,072 km (666 miles) and 1,120 km (696 miles) respectively. Rhea is about 50% larger again, 1,528 km (949 miles) across, then comes giant Titan, and the smaller outer moons Hyperion and Iapetus. The pattern in the sizes of moons is quite similar to the pattern among planets in the Solar System, with small inner moons, giants in the middle, and then smaller outer moons. So Saturn's satellites probably formed in a similar way to the Solar System itself (see p.206), with a great doughnut-shaped cloud of dust and ice colliding and coalescing into solid worlds. Large worlds formed where the cloud was at its thickest, while smaller moons were born on the thinner inside and outside edges.

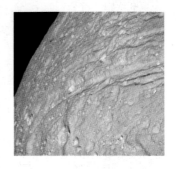

The southern end of Ithaca Chasma on Tethys peter out into a heavily cratered landscape, as revealed in this orbital photo.

Tethys

From a distance, Tethys looks bright and icy, like a larger version of Enceladus. But as you get closer, you'll start to see differences. It's far more heavily cratered, its leading face is dominated by a huge basin called Odysseus (all the names on Tethys come from Homer's epic poem *The Odyssey*, by the way). Despite being about 400 km (248 miles) wide, Odysseus is surprisingly shallow. The icy crust has flowed and shifted since the crater formed, leaving it a ghost of its former self.

Tethys orbits 294,700 km (183,100 miles) from Saturn, locked like most moons with one face permanently turned to its parent planet. While the outward side has little to

Odysseus dominates one face of heavily cratered Tethys.

recommend it apart from craters, the Saturn-facing side is dominated by a huge canyon system called Ithaca Chasma. Up to 4 km (2.5 miles) deep in places, the trench cuts a broad swathe across the cratered northern hemisphere. Ithaca Chasma may be halfway around the moon from Odysseus, but it runs parallel to the crater wall, and that's unlikely to be a coincidence. It's most likely that this great fault was created during the Odysseus impact itself, or when the crust later shifted to flatten out the crater.

The looming sphere of Saturn rises over Ithaca Chasma on Tethys.

Dione

Tethys's near-twin sits further out, at an average distance of 377,000 km (234,000 miles) from Saturn. It's a world of two halves, with a brighter hemisphere covered in smaller craters, and a darker half that bears the scars of heavier impacts. As on Tethys, a lot of the craters have 'slumped' and flattened out as the icy crust has shifted over long periods of time.

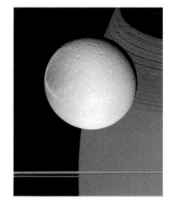

Dione hangs in front of Saturn, just above its ring plane. Note the dark patches and bright streaks on one side of the moon.

Catching up with the moon in its orbit, you'll get a good look at Dione's most intriguing feature, once known by the picturesque name of 'wispy terrain'.

At first, it looks as if Dione's dark regions are covered in patches of frost, streaking the surface. Closer in, though, the streaks reveal their secret: they are actually long cliff faces, bright with smooth, exposed ice, and running in parallel swathes across the landscape. What they lack in height, they make up for in number, since hundreds of these streaks run across the trailing face. They're strong evidence that Dione's crust once began to shift and crack apart, perhaps pulled by tidal forces earlier in its history.

Narrow, steep ice cliffs run across areas where Dione's crust was stretched and cracked in its distant past.

Rhea

Because Rhea is bigger than Tethys and Dione, you might expect lots of evidence for icy volcanoes here, and perhaps more tectonic activity. Unfortunately, Rhea turns out to be a

ICY ERUPTIONS

The cratered plains of Tethys and Dione both show a lot of variation. Some regions are scarred with many large impacts, while others have a mere scattering of relatively small craters. This is all a dead giveaway that parts of the planet have resurfaced themselves, probably at around the same time our own Moon was pulling the same trick, towards the end of the heavy bombardment about 3.9 billion years ago (see p.206).

There's a difference from the Moon and the terrestrial planets, though. These are icy moons, and the stuff that erupted onto their surfaces was clearly a mixture of rock and ice. But whoever heard of a volcano erupting ice?

Of course, it turns out there is a way, and it relies on the presence of another chemical, ammonia, in Saturn's icy moons. Fortunately we already know that there's ammonia hanging around in the cold of the outer Solar System, so it's not too much of a leap to believe there might be some on Saturn's moons too.

When ammonia mixes with water at very low temperatures, it forms a solution that has some very weird properties. For one thing, it's got a much lower freezing point than pure water. For another, when cooled close to its freezing point it becomes a viscous, oozing fluid that behaves in almost exactly the same way as lava. So it seems that early in their history, Tethys and Dione were warm enough to melt the water-ammonia mix, allowing it to erupt and ooze from the interior onto the surface. In places, it wiped away the record of previous cratering, leaving a blank slate, the icy equivalent of the Moon's freshly solidified seas.

bit of a damp (or rather, deep-frozen) squib. You'll soon notice that it's heavily cratered and darker than either of its inner neighbours, sure signs that its surface is very ancient. The only signs of activity it can offer are some ice cliffs similar to Dione's, showing that the crust shifted a long time ago. Unless you're keen on craters, you'll probably move on to the enticements of Titan with barely a backward glance.

But why is Rhea so different, and such a let-down? The best bet is that its larger size and mass allowed it to freeze solid in a way the smaller moons never could. Apparently (say the scientists) there are two forms of ice. The stuff you pull out of the icemaker isn't too different from the stuff that makes up glaciers, or Saturn's smaller moons. It's comparatively lively, and able to slip and shift around. The stuff inside Rhea, however, is 'Ice II', a no-nonsense form that's had all the fun squeezed out of it. In the deep cold of Rhea's interior, it won't even play with ammonia (see 'Icy eruptions'), so there was no way for any cryovolcanism to get going and inject some life into the old iceball.

Rhea displays its heavily cratered surface and some of its Dione-like ice cliffs.

Titan

The cloud-covered and mysterious world of Titan is almost as big a draw for Saturnian tourists as the rings. First discovered by Christiaan Huygens (the same chap who figured out the true shape of the rings) in 1655, Titan is one of the biggest moons in the Solar System. In fact, at 5,150 km (3,200 miles) across, it's bigger than both Pluto and Mercury.

Most of the reasons for people getting excited about Titan lie on the surface, and to see them close-up, you'll have to get through the atmosphere first. Methane smog blocks out the view from orbit, although with the right filters you can pick up some surface detail, mostly light and dark splodges.

Descent through Titan's atmosphere is no big deal – this far out in the Solar System, the temperatures are low and the winds fairly sluggish. The air itself is mostly nitrogen, with a large amount of methane which gives the distinct orange colour. The clouds, too, are formed by methane droplets condensing in the atmosphere.

TITAN'S VEIL

Why does Titan, of all the Solar System's moons, have such a thick atmosphere? Like Earth, Titan is a Goldilocks planet – albeit one where Goldilocks' porridge would instantly freeze into a solid oaty mass around her spoon. Closer to the Sun, Titan would be warmer, and its gravity too weak to hold an atmosphere. Further out, gas would stay frozen on the surface. But here, it's just cold enough for gas atoms to move around sluggishly, never quite getting up the energy to escape into space.

From orbit above Titan, the dense orange atmosphere seems full of mystery and foreboding.

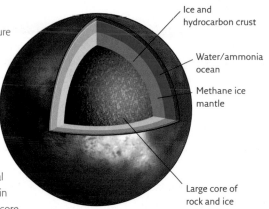

INSIDE TITAN

Titan seems to have a unique internal structure to match its unique outward appearance. Beneath its crust of organic chemicals, it's thought to be an icy moon similar to its neighbours, with a large core of rock and ice. However, directly beneath the surface lies an ocean of water and ammonia, just the thing to power cryovolcanoes and help reshape the surface. This layer has been heated up several times in Titan's history by chemical changes in the methane ice mantle directly around the core.

Ice and hydrocarbon crust

Water/ammonia ocean

Methane ice mantle

Large core of rock and ice

Most of the time, Titan's cloud base is a few kilometres up. Below this, the skies are fairly clear, and you should start to get your first hazy views of the surface.

Titan's surface can be a shock after months or years away from home. It's quite a jolt to see the obvious traces of rivers, lakes and shorelines this far out in the Solar System, and enough to make you pine for Earth. But tough – if you've come this far, you're not going to go back without exploring, are you?

Unless you want your first steps on Titan to make you look like a novice skater at the Rockefeller Center, we recommend you mind where you park. Large regions of Titan's surface are covered in slicks of frozen methane and other organic chemicals (don't get the wrong idea when we call them organic, by the way – it's just chemist-speak for any carbon-based compound). If you take our advice, you'll aim for one of the dry, shallow lake beds. Just check it hasn't suddenly filled up since your charts were compiled, or you could find yourself paddling through liquid methane to the shore.

Head for the delta where one of the river channels empties into the lakebed, and you'll find a treasure trove of rocks and

FLAMMABLE TITAN

With all these hydrocarbon compounds around, Titan should be highly flammable. In fact, the only reason that it's not is the lack of oxygen, a must to make these chemicals burn. Since you'll be carrying a backpack full of oxygen and electronics into this lot, we suggest you take the same care that you would when organizing a fireworks display near an oil refinery.

other material washed down from the higher ground. The size of the rocks carried by the rivers suggests that their currents can have some force, although so far they've never been seen in action. The rocks themselves are lightweight compared to those of Earth or the terrestrial planets, however, because there's a lot of ice mixed into them. Despite its atmosphere and hydrocarbon coating, Titan's not so different from its neighbours in the Saturnian system.

If you're lucky, you may catch a light shower – Titan's bizarre weather causes methane to rain from the skies, and it's possible that large areas of the planet were flooded not too long ago. In fact, the average surface temperature of around -180 °C (-290 °F) is just right for Titan to have developed a full-blown 'methane cycle' similar to the water cycle that transforms water on Earth between ice, liquid, and vapour. So methane falls from the sky as rain, washes into rivers and accumulates in lakes, freezes into ice, and evaporates back into the atmosphere. This means that Titan's landscape is on the receiving end of a lot of erosion, explaining its eerie similarities to Earth.

One big question, though, is where all the methane comes from in the first place. Methane in any planet's

Icy boulders lie around in the dried-up river delta where the Huygens lander first touched down on Titan's surface in 2005.

As the clouds part, a coast-like landscape of rivers, bays and islets appears below, in the midst of a 'sea' of sand dunes.

atmosphere gradually breaks down when it's exposed to sunlight, and there's been more than enough time in Titan's long history for the gas to have disappeared completely. Something must be renewing it periodically, and Titan experts think they know what it is. They believe that the moon has a thick crust of water and methane ice on top of a mantle of ammonia-water mixture (the same kind of mix that resurfaced parts of Tethys and Dione, see 'Icy eruptions', p.157). Titan's interior is still slowly cooling down from its formation, and when the mantle began to freeze about half a billion years ago, it released excess heat, warming the crust and triggering outbursts of icy cryovolcanism, pumping methane into the atmosphere.

It's possible that eruptions are still occasionally happening today, though they will have slowed down a lot. Certainly there are 'hot spots' on Titan's surface that look as if they could be the result of erupted ice cooling down.

Unfortunately Titan's smoggy atmosphere on Titan never clears enough to reveal Saturn hanging in its skies.

Sponge-like Hyperion must count as one of the Solar System's strangest and prettiest worlds.

Hyperion

Orbiting just beyond Titan, Hyperion is an oddity. For one thing, it's in a distinctly elliptical orbit, with an average distance of 1.48 million km (919.600 miles) from Saturn. That will make your orbital transfer a little more tricky, but as you draw in close to this strange moon, we think you'll agree it was worth the effort.

Hyperion is a bizarre-looking world: in places it looks more like a coral or sponge than a moon. It's almost as large as Mimas, but has an uneven shape, 330 by 260 by 215 km (205 by 161 by 134 miles). At that size, it should have been quite capable of pulling itself into a sphere, and as you draw close enough to take in details, it's pretty obvious what happened to it. At some point in its past, Hyperion took a direct hit from another large object, and shattered to pieces.

What survives is scarred almost beyond recognition. The most spectacular features are craters that look like water

CAVING ON HYPERION

For the more intrepid tourist, Hyperion's spongy surface offers another intriguing activity: some of the deeper craters offer access to the moon's equally porous interior. Potholing on another world needs a great deal of care, especially when the rocks are as jagged and sharp as Hyperion's, but the few who have done it (and come back alive) say it's an unforgettable experience.

drops frozen in mid-splash: something seems to have nibbled at Hyperion's surface, giving it a moth-eaten appearance

Hyperion also has a unique style of rotation. It spins in a completely chaotic way, with no obvious axis of rotation, and no measurable rotation period. The elliptical orbit and irregular shape are most likely to blame for this awkwardness.

Elsewhere in the Solar System, moons and asteroids shattered in collisions are often able to pull themselves back together, even if their gravity is weak. For Hyperion, though, that wasn't an option. The gravity of nearby Titan plucked at the smaller fragments flung off in the impact, pulling them into its atmosphere and onto its surface. A fair number missed Titan completely. These bombarded the surface of the next moon in, Rhea, where their scars form a distinctive family of impact craters, all around 20 km (13 miles) across, and all formed at about the same time. Eventually, the larger fragments of Hyperion pulled themselves back together, following a new, disrupted path around Saturn.

Saturn peeks over the razor-sharp crags that surround one of Hyperion's major craters.

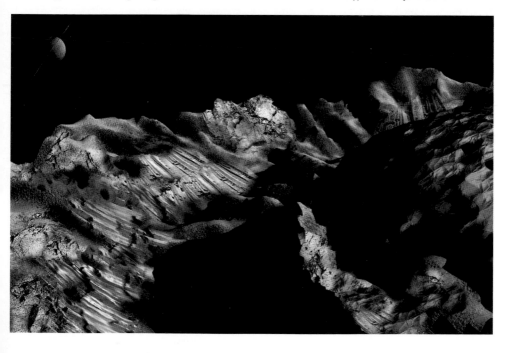

Iapetus
(and some bits of Phoebe?)

Iapetus (another hard-to-pronounce Saturnian moon – say it 'Eye-app-it-us') is the outermost of Saturn's large moons, a near-twin to Rhea at 1,436 km (892 miles) across. Plenty of other satellites orbit further out than its 3.56 million km (2.21 million miles), but these are just assorted bits of captured debris, mostly stray asteroids or their icy equivalents, the centaurs (see p.167).

You should be able to tell that something's up with Iapetus long before you can see any surface features. During your first approach to Saturn, keep an eye on the dance of its satellites. You'll soon spot that Iapetus changes its brightness considerably and, depending on your angle of approach and timing, may even disappear completely as it rounds one side of the planet.

Because Iapetus is tidally locked with one face towards Saturn, it's pretty easy to work out that one half of the moon is much darker than the other, and as you get close, you'll spot the obvious difference. Iapetus looks as if a graffiti artist attacked it with a can of spray paint, but only finished one hemisphere before the police arrived. The interesting question, of course, is what really happened. As you slip into orbit, you'll be able to see some details. Both terrains seem equally cratered, and there are places close to their boundary where craters in the dark, reddish material have created light spots, a sure sign that Iapetus is a light world with a dark coating, and not the other way round.

The best explanation for the moon's strange appearance is that its next-door neighbour is to blame. This is dark and irregular Phoebe. Ever since the Cassini spaceprobe flew past it on its way into orbit around Saturn way back in 2004, most people have agreed that Phoebe is a captured world, a large comet nucleus or a centaur, 230 km (143 miles) across. It's the largest of Saturn's irregular moons by a long way,

As Iapetus orbits one way around Saturn, does it fly into dark material blasted away from Phoebe?

and circles Saturn in the opposite direction to the 'normal' moons – a sure sign that it was captured in a random close encounter, rather than forming in its current orbit.

What seems to happen is this: Phoebe, like all the other moons, gets bombarded with tiny micrometeorites as it circles Saturn. These create a trail of 'soot' that gradually spirals in towards the planet, and as Iapetus ploughs through Phoebe's exhaust fumes, its leading hemisphere gets coated in muck.

All very neat, but there's just one problem. Phoebe's surface is dark grey in colour, while Iapetus's dark half is distinctly red. If Phoebe dust is really to blame, then something must be changing it when it gets to its destination. One idea is that the heat produced as the dust hits Iapetus is enough to melt methane in the moon's crust. Dust and methane then react to form tarry reddish compounds. As a result, a walk on the dark half of Iapetus can be a messy business, rather like crossing a freshly surfaced road on the hottest day of the year.

The dark, misshapen outer moon Phoebe is a captured object, and probably the source of Iapetus's dirty coating.

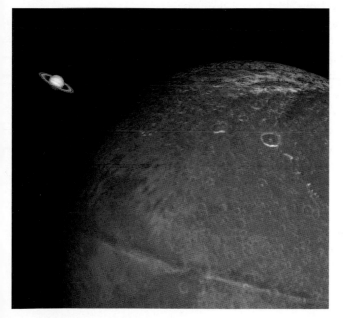

Saturn hovers in the distance in this view over Iapetus's dark half. Parts of the brighter terrain are just visible near the north pole. Note also the distinct ridge around Iapetus's equator.

Further afield – Chiron

A variety of interplanetary strays loiter around the orbits of Saturn and the other outer giants. Known collectively as centaurs, they're quite distinct from the asteroids of the inner Solar System. The first to be discovered, the largest, and still the best known, is Chiron, which goes round the Sun once every 51 years, coming within Saturn's orbit at its perihelion, but getting almost to Uranus at its most remote. If you're in the vicinity anyway and the orbits happen to work out, it's well worth dropping in on a centaur such as Chiron or Pholus.

Centaurs have pretty dark surfaces, so there's an art to spotting one before you're right on top of it. Infrared imaging will help out here – centaurs might be deep-frozen, but they'll still shine like beacons compared to the cold of surrounding deep space. If you're heading to Chiron or one of the other relatively sunward centaurs, you may also notice that your destination is looking a little fuzzy round the edges. It's not an equipment fault, but a hint that some centaurs have thin atmospheres of their own.

Slipping into orbit alongside Chiron, it's easy to see why. This 180-km (110-mile) chunk of rock and ice looks just like a giant comet, with plumes of vapour puffing quite gently from cracks in its surface and enfolding the entire body in a thin envelope of gas and ice particles.

Centaurs, it seems, are 'ice dwarf' worlds that have wandered in from the Kuiper Belt beyond Neptune (see 'The Outer Limits', pp.190–203). They probably had their original orbits disturbed by close encounters with Neptune, but stayed far enough away to avoid the destructive tug of Jupiter, that would either turn them into true monster comets, or just as likely break them up or fling them from the Solar System altogether. They can't have loitered in this area for very long in astronomical terms – crossing the orbits of giant planets is a dangerous business, and Saturn's gravity is still strong enough that it will inevitably send Chiron tumbling towards the inner Solar System, perhaps within a few thousand years.

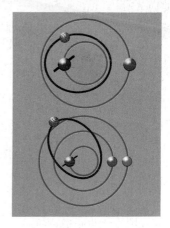

The orbits of Chiron (top) and Pholus (bottom) shown in relation to those of Saturn, Uranus and Neptune.

WHAT'S IN A NAME?

Try not to confuse Chiron with Charon, Pluto's largest moon. Chiron is, appropriately enough, named after a centaur from Greek mythology – one of a breed of wise, half-man, half-horse warriors with a dislike of saddles (and presumably a weakness for polo mints). Charon is an altogether more miserable character, the boatman who carried dead souls over the River Styx into the underworld ruled by Pluto.

At its closest approaches to Saturn and the Sun, Chiron's surface activity is at its strongest.

NEPTUNE
IT'S A BLUE THING

URANUS

THE GAS IS ALWAYS GREENER ON...

Between the ice giants

Uranus and **Neptune**, the Solar System's outer 'ice giants', make an ideal two-centre holiday. Plan your trip to Uranus well in advance – its bizarre seasons make it look dull and inactive for much of its 84-year orbit. Neptune is a more reliable source of activity, a blue world circled by dark storms and home to some of the highest winds in the Solar System. Even if the planets are quiet during your visit, there are many highlights among their moons and rings, including Miranda, possibly the most bizarre world in the entire Solar System, and Triton, Neptune's giant moon with its active ice geysers.

Even these days, the Solar System beyond Saturn is still the preserve of the serious traveller. Journey times are so long that you're looking at several years away from home, something that only career-break city slickers, wealthy retirees and a few trustafarian students can contemplate.

Getting there

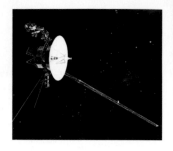

If you're going to do it, you'll want to do it well, and if you're going as far as Uranus, you might as well double up the experience by carrying on to Neptune. The ice giants may be near-twins at first glance, but they display many differences, and Neptune's big moon Triton will give you a glimpse of the exotic Kuiper Belt without an even longer journey.

The worlds beyond Saturn were just indistinct blobs until the late twentieth century, when a single long-distance spaceprobe, Voyager 2, visited both Uranus and Neptune.

Uranus takes 84 years to orbit the Sun, and Neptune 164 years, so you probably won't have much opportunity to pick your travel conditions. However, it's worth bearing in mind that the fastest way of reaching Neptune is with a gravitational slingshot at Uranus, which has implications for how you want the two planets to be lined up (see p.15).

Weather on Uranus is highly seasonal, so you'll really want to avoid the dull periods that coincide with midsummer and midwinter for each hemisphere. Way back in 1986, Voyager 2, the first spaceprobe to fly past Uranus, arrived around midwinter, and found Uranus in one of these regular sulks, its globe dull and almost entirely featureless. To avoid making the same mistake, try to visit in spring or autumn, when the weather systems really come alive and Uranus can give the other giants a run for their money.

GAS GIANTS OR ICE GIANTS?

You'll notice we call Uranus and Neptune 'ice giants' rather than gas giants. There's a good reason for this, and it's not just that they're cold. While Jupiter and Saturn are almost entirely hydrogen gas, Uranus and Neptune are made from a mix of other materials, 'ices' in the chemical sense (meaning chemicals with low boiling points, which can exist as frozen ice or vapour depending on their distance from the Sun and their temperature, and as liquids on the surface of planets or within them). Beneath their outer atmospheres, both planets have interiors that are mostly liquid or slushy ices – a mix of water, methane, ammonia, and other chemicals.

Uranus

Although the ice giants are very similar in size and colour, Uranus is the slightly larger and greener of the pair. As it looms out of the darkness in the months before your closest approach, you should be able to track the positions of its satellites. Almost as soon as they had discovered the major moons, astronomers realised that there was something strange going on at Uranus. You may want to work it out for yourself, or you may prefer to cheat and take a peak through binoculars or a telescope. This will reveal the faint system of narrow rings that encircles the planet itself, and give the game away. Uranus looks like a giant bullseye

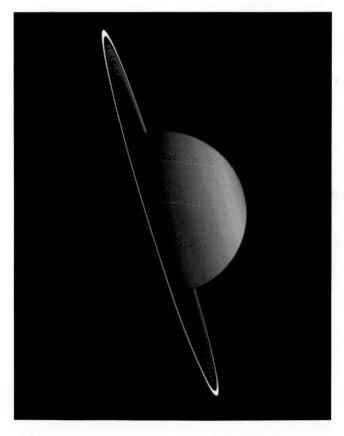

The tilted rings around Uranus clearly show how the entire Uranian system is tipped over by more than 90 degrees.

because the planet, its moons, and its rings, are all tipped over onto one side.

Uranus can't possibly have formed this way up, so it's clearly been knocked over at some point in its history, most likely the victim of a cosmic head-on crash with a large planetoid wandering through the outer Solar System. Whatever happened, it must have been long, long ago, since Uranus's satellites have had plenty of time to stabilise in orbits around the tipped-over equator, and there's no sign that the system has suffered a recent trauma.

There's also no sign of 'the other guy' in this interplanetary collision: it was probably destroyed entirely, or flung completely out of the Solar System. And because Uranus was a soft target, there's not even a big crater to admire. However, the planet does bear the scars of its violent past, in the form of its bizarre seasons. Tipped over at 98 degrees from the vertical, it suffers much more extreme seasons than the other planets. Only a narrow band around the equator gets to enjoy a permanent daily sunrise and sunset matching the planet's 17-hour rotation period. The poles themselves have to put up with days and nights each lasting 42 years, as first one pole, and then the other, takes its turn to face the Sun. Mid-latitudes get a mix of both patterns. They spend some of each Uranian year in permanent darkness, some with a daily sunrise and sunset, and some in permanent sunlight.

Viewing Uranus 'the right way up', you can start to appreciate just how weird the planet's seasons are. Here, the northern hemisphere is experiencing its long summer.

INSIDE URANUS

Uranus's outer atmosphere goes down to a depth of about 7,250 km (4,500 miles), the pressure and temperature gradually increasing until it merges with the liquid mantle. There's no sharp boundary between the two – the atmosphere just gets thicker and wetter until it has changed from gas to liquid. The mantle is largely a slushy mix of methane, ammonia and water ice, and at its centre lies a core of rock and ice, probably about the size of Mars.

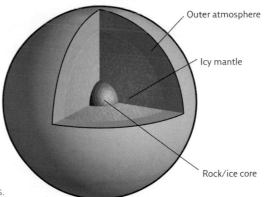

Outer atmosphere

Icy mantle

Rock/ice core

EXPANDING THE SOLAR SYSTEM

It's hard to believe now, but for most of recorded history, astronomers thought the Solar System stopped at Saturn. It was a huge surprise, therefore, when in 1781 a German musician and amateur astronomer called William Herschel spotted a fuzzy, dark object that he thought was a comet. Herschel lived and worked in England, and built his own telescopes. As he tracked the 'comet' over several months, he realised that it was far too slow-moving for a nearby object. Instead it must be huge, and far beyond the orbit of Saturn: in other words, a new planet. The discovery made Herschel so famous that today he'd have been on the cover of every issue of *Heat* magazine for months. Quite a lot of people even wanted to name the planet after him, but Herschel himself shrewdly named it 'George's star', after King George III (a gesture that got him a cushy job as the King's personal astronomer). The name Uranus (Saturn's father in Greek myth) was suggested by German astronomer Johann Elert Bode.

All this makes the Uranian weather a meteorologist's nightmare. Around midsummer/midwinter, when the temperature difference between the poles is at its greatest, huge currents build up to transfer heat from one hemisphere to the other. These cut across the general air circulation caused by the planet's rotation (the same 'coriolis' force which usually generates currents from west to east parallel to a planet's equator). In the ensuing clash, neither current wins out, and the upper layers of the Uranian atmosphere become an apparently placid, featureless millpond.

As Uranus continues its slow trek around the Sun, approaching the equinox (when each hemisphere receives equal sunlight), the temperature differences even out and the currents driven by them die away. The coriolis forces now become dominant, a pattern of bands around the planet establishes itself, and storms and other features begin to form. This is why it's always best to plan your visit (if at all possible) around the Uranian seasons. Just a few years can make all the difference to whether you see a featureless turquoise ball, or a stormy, active world.

The rings

The Uranian ring system is well worth seeing on your whistlestop tour. Although it can't compete with Saturn for sheer spectacle, it's the very difference between the two systems that is so fascinating. While Saturn has countless rings so close together that they merge into great planes, Uranus has fewer than a dozen very thin and thread-like rings, each entirely separate from the others.

The rings are also made of surprisingly dark material, much less reflective than the chunks of shiny iceberg orbiting Saturn. Combine the darkness of the rings with the rapidly dwindling light from the Sun, and you'll be lucky to see them against the blackness of space. Your best bet is to pick the rings up on radar or get them silhouetted against

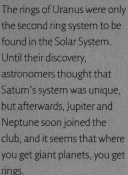

FINDING THE RINGS

The rings of Uranus were only the second ring system to be found in the Solar System. Until their discovery, astronomers thought that Saturn's system was unique, but afterwards, Jupiter and Neptune soon joined the club, and it seems that where you get giant planets, you get rings.

Coming a couple of decades before Earth-based telescopes grew powerful enough to photograph them directly, the discovery itself was sheer good luck. Way back in 1977, astronomers aboard NASA's flying observatory were gearing up to record one of those rare occasions when Uranus passed in front of a star. To their surprise, they saw the star wink on and off several times before it disappeared behind the planet, and then do a repeat performance after its reappearance. They soon worked out that these dips revealed a set of nine rings, and when Voyager 2 flew past the planet in 1986, it added two more to the collection.

the planet. This should reveal that they're chains of surprisingly uniform boulders, around 1 metre (3 ft) across. While Saturn's rings are made of water ice, Uranus's ring fragments turn out to be made from methane ice, which is probably why they're less reflective.

Slip around the outermost 'Epsilon' ring, and you'll eventually come to the small moons Cordelia (on the inside edge) and Ophelia (on the outside edge). Both are about 40 km (25 miles) across, and their gravitational influence helps to keeps the Epsilon ring in line. According to one theory, these 'shepherd moons' may even be crumbling to pieces and supplying new material to the rings, but no-one's found the clinching evidence yet.

This profile of the entire ring system shows differences in the brightness and colour of the various rings, caused by their different compositions. The upper (outer) Epsilon Ring is by far the brightest.

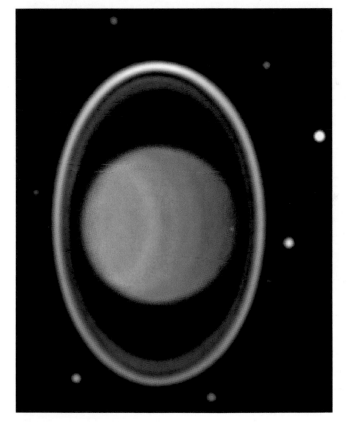

An Earth-based, enhanced photo shows the major moons of Uranus circling the planet and its rings. Note also the bright storms in the far hemisphere.

Miranda

This tiny moon, just 480 km (300 miles) across, is probably *the* tourist attraction of the Uranian system, a mishmash of a world whose traumatic tale is written across its chaotic landscapes.

As you fly in towards close approach, the first thing you'll spot will probably be the huge 'racetrack' features called coronae. The human eye is brilliant when it comes to spotting patterns, and these bizarre concentric features look like nothing so much as a Grand Prix circuit.

Next thing to look for, we'd suggest, are the cliffs in the southern hemisphere, enormous scarps in the moon's surface that tower so high that they cast long shadows at any time of day.

As you get closer in, new features reveal themselves thick and fast. Some areas are heavily cratered, and so probably haven't changed since early in Miranda's history. Others are smoother, bright and less cratered – clearly they were wiped clean of impacts at some point in the moon's distant past. The coronae themselves turn out to be surprisingly varied close up. Although they are generally darker than the rest of the terrain, some have bright patches within them. The tracks also have different patterns: some have smoothly curving corners, while others, most obviously Inverness Corona (nicknamed 'The Chevron'), make sharp, sudden turns. In general, Miranda looks as if it was slung together by an overenthusiastic planetary architect who couldn't bring himself to leave anything out.

It's a bizarre little world, for sure, and was quite a puzzle for the late-twentieth-century scientists faced with explaining it from Voyager 2's pioneering handful of images.

Their first thought was that Miranda was a genuine planetary Frankenstein, perhaps the victim of an ancient collision that had pulled itself back together through gravity. On reflection, though, most decided that Miranda

A global view of Miranda shows its distinctive patchwork appearance.

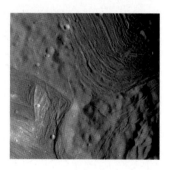

Even from orbit, regions of Miranda's surface still look like puzzle parts jammed back together by an enormous and impatient toddler.

probably never went to pieces completely. Instead, it seems more likely that the moon's interior melted and its crust cracked, allowing large chunks of the surface to slip beneath the molten mush, never to be seen again. Different processes created new features in the freshly healed surface, before another bout of heating and melting churned the whole surface over again.

Miranda could have been kept warm through its own internal heat, and the energy that came from large meteor impacts. Tidal heating (see p.124) may also have played a role – perhaps Miranda was warmed up during that turbulent period after Uranus was knocked from its axis, and the whole system had to rearrange itself in a new plane of rotation.

A flight across the coronae reveals that they probably formed as sections of the surface pulled apart from one another, and new material welled up to fill the gaps.

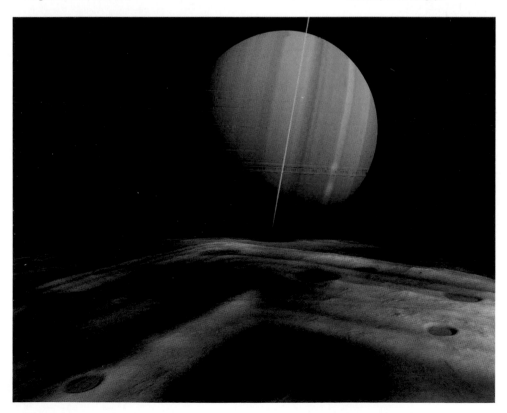

The outer moons

After Miranda, the other Uranian moons are a bit more restrained. Most travellers, committed to a swift flyby at Uranus so they can continue to Neptune, will only pay these worlds a flying visit, but each has its own story to tell.

You may have noticed elsewhere in the Solar System that moons tend to come in pairs, and the same rule holds true at Uranus. The next substantial moons out from Miranda are Ariel and Umbriel, both about 1,170 km (730 miles) across. Then come Titania and Oberon, both roughly 1,550 km (960 miles) wide.

Ariel and Umbriel

All the larger Uranian moons have relatively dark surfaces, but Ariel's is the brightest. As you zip past, you should have a chance to admire the deep canyons that run across the surface. The braided, parallel trenches of Kachina Chasmata are particularly beautiful. You might also notice that Ariel's surface is quite fresh and the craters peppered across it are relatively small, which suggests that a lot of the moon was resurfaced at some point in its history. You may even spot where the canyon floors have filled with bright, icy material that obviously welled up from inside the moon. As the

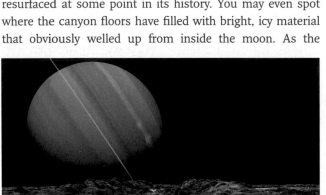

Ariel's cracked surface is scattered with relatively few, small craters.

A view down Kachina Chasmata towards the rising sphere of Uranus. Note the icy canyon floor.

second closest major moon to Uranus itself, Ariel must have received a lot of tidal heating in its youth. It didn't melt all the way through like Miranda, but could certainly have warmed up enough for ammonia-powered cryovolcanoes to redecorate the surface (as seen on Saturn's moons Tethys and Dione, see p.157). As Ariel froze again, it expanded from within, cracking the surface and opening up the canyons.

Compared to Ariel, Umbriel is pretty boring. It's obviously darker, and has more craters than any of the other Uranian satellites. It's pretty clear that it's been deep-frozen since early in its history, acting as little more than target practice for passing meteoroids and comets.

Umbriel bears the scars from countless impacts. The bright spot is a puzzle – since this view is from the pole, it actually lies on the equator.

Titania and Oberon

Next up is Titania, largest of the Uranian moons. Even on a flying visit, it looks like Ariel's big sister – the surface isn't quite so bright, but it's got a fairly similar mix of cratered plains and deep scars. Titania would never have received the same kind of tidal heating as Ariel, but it's big enough not to need it, and, along with Oberon, is unusually rocky for an outer satellite. This means that heat retained from its formation, together with energy generated by radioactive elements within, would have been more than enough to power its activity. It created the faulting that split Titania's surface, and may even have resurfaced some parts of it at a very early time.

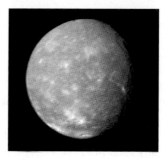

Titania is one of the brighter satellites of Uranus, with canyons and resurfaced regions resembling Ariel.

Last stop on your way out from Uranus is Oberon. Broadly similar to Titania, it has fewer obvious trenches and more large, ancient craters. Both these differences suggest that Oberon has not had such an active past as Titania. However, it makes up for this with some unique features. Several of the largest craters are surrounded by bright splashes of ejecta, while their floors are covered in darker material. It seems that these impacts threw out bright, reflective ice from the upper layers of the planet, but also allowed dark, carbon-rich methane ices to well up from inside Oberon's mantle and flood the crater floor.

A view of Oberon shows its unusual dark-floored craters. Can you spot the huge mountain on the limb?

Neptune

Considering how far it is from the Sun, you'd expect Neptune to be about as much fun as a rogue fish finger that's been buried at the back of your freezer for several years. The fact that both the planet and its major moon, Triton, are fascinating, active worlds is something that astronomers are still coming to terms with. It's also something that makes them a major draw for travellers on the Solar System's outer rim.

Neptune is slightly smaller than Uranus, receives a lot less sunlight, and has a naturally darker colour – all of which mean you'll be that much closer to it before it appears out of the gloom. However, by the time it's properly visible to the naked eye you should already be able to spot surface details if you know where to point your binoculars. Dark spots are the most obvious feature, great oval storms with the same cause as the spots on Jupiter and Saturn. The first space probe to get this far out, Voyager 2, happened to zoom past when a huge spot, immediately named the 'Great Dark Spot' (GDS), dominated one entire hemisphere of the planet. It turned out not to be so great after all, since when astronomers looked again a few years later, the GDS had disappeared.

But Neptune's storms are always coming and going, and you'll be unlucky not to see something impressive. At times, different weather zones wrap themselves around the planet, creating cloud bands very similar to those on the inner giants.

THE NEPTUNE AFFAIR

The row over Neptune's discovery was the astronomical scrap to end them all. A few short decades after Herschel's chance discovery of Uranus, most astronomers admitted that something was up with the new planet's orbit – most likely the effect of another unseen world. Around 1843, English maths wizard John Couch Adams worked out where the planet was, but it seems he didn't know anyone with a telescope, and the Astronomer Royal wasn't interested. A couple of years later, wily Frenchman Urbain Le Verrier did the sums independently, and got his pal Johann Galle of the Berlin Observatory to check the location. Galle found Neptune first time, on 23 September, 1846. One-nil to the French, clearly, but the Brits have been complaining about it, off and on, ever since.

Riding the Scooters

Bright scooter clouds hang above the deep ocean-blues of Neptune's lower atmosphere.

Neptune's atmosphere may look calmer than Jupiter's or Saturn's, but it's deceptive. The planet actually has the most powerful winds in the Solar System, and it's well worth taking a dip into the outer layers to enjoy the experience of 'sailing' alongside Neptune's clouds.

Neptune's winds aren't only driven by heat from the Sun. The planet itself generates twice as much energy as it receives, and this means that the cloudtop temperatures of Uranus and Neptune are roughly identical at around -215 °C (-350 °F), despite Neptune's much greater distance from the Sun. What's more, with a tilt similar to Earth's, Neptune doesn't suffer from any of the seasonal complications seen on Uranus. Instead, the coriolis forces created by Neptune's 16-hour rotation are in full control, streaming winds and cloud patterns around the planet.

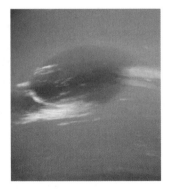

The winds are at their strongest in the uppermost layers of the atmosphere. Fly in alongside one of the small white 'scooters', and you'll be swept along at more than 2,000 km/h (1,250 mph). That's fast enough to carry you round Neptune's

White, high-altitude scooters often form near the dark storms of the lower atmosphere, though their faster speeds soon carry them away from their origins.

Inside Neptune

Neptune's a bit smaller than Uranus, and the structures of the two planets are generally similar. But the outer world is slightly denser than the inner one, and that's a clue that Neptune's core is slightly larger.

Neptune's internal energy source is probably gravitational contraction beneath the surface – the same as for Jupiter and Saturn. There's a spectacular difference here, though, caused by Neptune's icy composition. As denser materials sift down through the planet's mantle and onto the core, they create heat, and the heat breaks up methane ice in the mantle, allowing pure carbon to rain down through the layers. Where carbon atoms are forced up against each other in these high-pressure conditions, they join together to form the hardest substance known to man. Yes, inside Neptune it's probably rain diamonds!

Slushy ice mantle

Rock/ice core

Outer atmosphere

equator in a little over three Earth days. Elsewhere, long white streaks of cloud form where warm vapour rising from the planet's interior turns to droplets in the cold upper atmosphere. As the weak sunlight casts their shadows onto the solid dark blue of the lower cloud layers, it's hard not to recall clouds over the Earth's own Pacific Ocean.

Every good giant planet needs a ring system, but Neptune's rings are a pretty poor collection. There are just five narrow bands, dark and quite similar to the rings of Uranus. Tiny shepherd moons help keep them on the gravitational straight and narrow, but the weird thing is that they're all much thicker on one side of the planet than the other. Back when astronomers first went looking for Neptunian rings, using the same technique that had revealed Uranus's system (see p.174), they found that something was blocking out starlight on one side of Neptune, but not the other. For a while, it seemed that Neptune must have incomplete 'ring arcs' instead of full-blown rings, but the Voyager 2 flyby of 1989 revealed the truth. Only the thickest parts of each ring are dense enough to briefly block out starlight.

Neptune's thin rings only become visible when backlit by the Sun. Even then, the light of a crescent Neptune can almost drown them out.

Neptune's moons

Compared to the other giant planets, Neptune's satellite system is pretty sparse and unimpressive. A handful of shepherd moons orbit amongst the narrow rings, and there's a number of small captured moons much further out, but there are only three more substantial satellites – Proteus, Nereid and Triton.

Proteus is a roughly egg-shaped lump of rock 440 km (273 miles) long, with one comparatively huge impact basin 255 km (158 miles) across. It orbits just outside Neptune's ring system in less than 27 hours and, to judge by the amount of cratering on its surface, may have donated quite a lot of material to the rings.

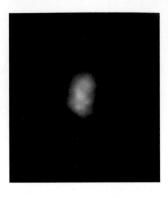

For most visitors, a distant, blurry view of outer Nereid is all they can expect to see.

Nereid is slightly smaller, 340 km (211 miles) across, with a highly eccentric orbit that takes it from 817,000 km (507,000 miles) above Neptune at closest approach, way out to 9.5 million km (5.9 million miles) at its most distant. It takes almost an Earth year to complete a single orbit, and unless you happen to be here when it's close in, it's probably not worth chasing after. Normally, you'd write off anything in an eccentric orbit as just another captured asteroid or centaur, but Nereid's different. The closeness of its approaches to Neptune suggest that it started out in a normal orbit and was then booted into its current ellipse.

Fortunately, Triton more than makes up for the shortcomings of its companions (it may also be responsible for them, but we'll come to that later...) It's the sexiest moon this side of Titan, with a bizarre surface, unexpected activity, and a traumatic history that gives Miranda a run for its money.

Triton

Assuming you're turning around at Neptune, it's worth taking the time to slip into a parking orbit above Triton and get a long, leisurely look at the blue and grey world below.

NAMING TRITON

Triton was the biggest discovery in the scientific career of William Lassell, a Liverpool brewer and amateur astronomer who made the finest instruments of the mid-nineteenth century. He spotted Triton barely a fortnight after Neptune itself had been discovered, but strangely he never suggested a name for the new moon, which remained anonymous until 1880, when Triton was suggested by French astronomer Camille Flammarion. Even then, the name was not officially accepted until around 1950.

The two major types of landscape are obvious even from a distance – relatively smooth greyish plains scarred with some craters, and the bizarre 'cantaloupe terrain', a blue-green-tinged surface that looks (unsurprisingly) like the skin of a cantaloupe or, if you prefer, like a sheet of bubble wrap with all its bubbles popped.

A little closer in, you may start to spot longish dark streaks across some of the grey plains, especially in Triton's southern hemisphere. This is where Triton starts to get really interesting. Once you're in a low orbit, take a look with binoculars towards the horizon in one of these streaked areas. A couple of seconds' travel along your orbit, and the parallax effect should make it clear that these streaks aren't on the surface at all – they're above it!

Forgive the exclamation mark, but in this case it's merited: that's how weird and unexpected Triton's activity is. Not only does it have these 'geysers', it's also got a thin but serviceable atmosphere for them to hang around in. Track a few of the streaks back to the vents where they billow out of the ground, and you'll make another connection. The streaks tend to run parallel to each other in the same direction, revealing that there are winds in Triton's atmosphere.

Even from a distance, Triton's clearly displays several distinct types of surface.

Neptune rises above the vast icy pits of Triton's frozen cantaloupe terrain.

Triton's a great place to get your feet back on solid ground, perhaps for the first time in years. It's a big moon, 2,707 km (1,681 miles) across, and with a surface gravity 1/12th of Earth's. That's just about right to get you reintroduced to the concepts of 'up' and 'down' without breaking any bones in the process. One thing you'll have to watch for, though, is the frost patches – parts of Triton are skating-rink slippery, so we recommend wearing ice boots.

To get the most out of Triton, try to land near the geysers of the southern hemisphere (the cantaloupe terrain looks attractively strange from orbit, but from ground level it's just a load of old hummocks). On the ground, you'll be able to walk across the sooty fallout from the vents, and take samples from the geysers themselves. It seems that most of the gas escaping from the geysers is nitrogen, while the dark soot is a mix of various carbon-based compounds.

If Triton's geysers work like the ones elsewhere in the Solar System, then liquid nitrogen underground is getting heated by contact with relatively warm rock or ice underground, then boiling away into space as soon as it finds a weakness in the crust. Well, we say 'heated', but it's worth remembering that nitrogen boils at just over -200 °C (-328 °F).

Geysers rise vertically from the icy terrain in Triton's southern hemisphere, before being swept up by winds in the thin atmosphere.

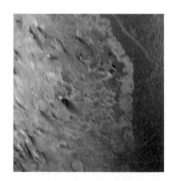

A photo from orbit reveals the shadows cast by geysers onto Triton's surface, and the dark streaks formed as dust and dirt drop out of the gas.

But even that temperature means Triton must have some kind of internal heat source. It's too small and too icy to still be holding heat from when it formed, so as usual, that leaves us with tides. Because Triton orbits Neptune in the 'wrong' direction, it suffers from bigger tidal forces than usual, and that seems to be enough to heat up the interior and power the geysers.

If our ideas about Triton's origins (see below) are true, then it would have been a lot warmer in its past, and that could explain a lot of its other weird surface features. If parts of the surface melted, that could also explain why the geyser terrain looks like a wrinkled 'skin' on top of the custard of the cantaloupe terrain. If the geyser landscape 'set' first, it could have floated on top of the rest of the planet as it cooled down and solidified after reaching its current orbit. The cantaloupe terrain itself, with its echoes of bubblewrap, might even have been shaped by pockets of lighter ices floating up from beneath it during these final stages.

TRITON'S ORIGINS

So how do we explain Triton, then? A great big lone moon alive with unexpected activity, in the middle of a satellite system that breaks the rules set by the other giant planets? The big clue lies in its orbit: a perfect circle, yes, but one that goes around Neptune the 'wrong' way compared to the planet's rotation. This is a sure sign that Triton doesn't belong here. It's a passing world that's been captured by gravity, just as surely as the smaller bodies orbiting the other giants once were.

Most likely, Triton started out as a Kuiper Belt Object (see p.197) – a world like Pluto, Xena, and the rest, that just happened to blunder into Neptune's path. Pulled at first into an eccentric orbit around the giant planet, tidal forces would soon have forced it into a more conformist, circular path. In the process, Triton would have played havoc with Neptune's original system of outer moons, in a series of close encounters that flung most of them out of orbit altogether. Only Proteus and the other inner moons survived unscathed, while Nereid got off relatively lightly, thrown into its present eccentric orbit.

When Triton arrived in Neptune's system (top), it found a collection of orderly moons. Most of these were ejected from the system altogether as Triton barged its way into orbit.

Further afield — Halley's Comet

The most famous of all comets only visits the inner Solar System once in the average lifetime, but if you've been unfortunate enough to miss it, or are just too impatient to wait, a trip to the outer edge of the Solar System is an ideal chance to catch up with the comet called Halley.

The comet's 1910 visit to the inner Solar System was one of the most spectacular of recent times.

This 15-km (9-mile) chunk of ice and rock orbits the Solar System in just the right situation to be a spectacular occasional visitor. Circling the Sun every 76 years, its appearances are frequent enough to be remembered from generation to generation, but not so common that it's yet burnt off all its icy goodness and become a cometary non-event. During each orbit, it spends just a few months sunward of Mars – far more time is spent around the orbits of Uranus and Neptune, and its aphelion point, furthest from the Sun, is between those of Neptune and Pluto.

Catching up with Halley in the outer Solar System, though, you may be left wondering what all the fuss is about. This far from the Sun, the comet is well and truly dormant, with only occasional cracks lined with fresh ice

Halley's dormant nucleus glints in the light of the distant Sun, during its long slow drift past Neptune.

breaking up its generally dark, crusty surface. It's hard to believe that, during its short visits to the Sun, it can develop a broad coma with the diameter of Jupiter, and a tail millions of kilometres long. Now all the material it shed has fallen behind the comet itself, smeared out into a long stream of particles that follows its orbit around the Sun. When Earth crosses near Halley's orbit in May and October each year, these are the particles that create the Eta Aquarid and Orionid meteor showers.

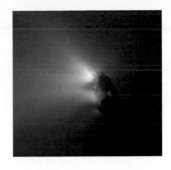

The first image of the nucleus was captured in 1986 by the European space probe Giotto.

Halley's greatest hits

Comet Halley is bright enough to get itself noticed on most returns, and it's been recorded throughout history.

240 BC: A bright appearance by Halley is recorded by Babylonian astronomers.

837 AD: The comet makes its closest ever approach to Earth, passing just 5 million km (3.2 million miles) away.

1066: Halley hurtles through Earth's skies and is taken as a bad omen for King Harold of England, who duly gets an arrow in the eye from Duke William of Normandy a few months later.

1301: Italian artist Giotto di Bondone uses the comet as a model for the Star of Bethlehem while painting *The Adoration of the Magi*.

1456: Pope Calixtus III excommunicates the comet, fearing it will have a bad influence over the Christian soldiers defending Belgrade from the Ottoman army.

1758: After a run-in with Jupiter, Halley turns up nearly two years late for its big date with history. In 1705, English astronomer Edmond Halley had suggested that the comets of 1607 and 1682 were actually one and the same, and predicted a return in 1757. The comet is given Halley's name, although he's not around to appreciate it.

1910: Halley makes one of its most impressive returns, sparking mass panics that Earth will be poisoned by cyanide gas from its tail.

1986: A bit of a damp squib appearance, in which Halley keeps its distance from Earth while at its brightest, is made up for by an armada of spaceprobes that return the first scientific data from the comet, including pictures of its nucleus.

It might not look much like a comet to you or me, but according to the embroiderers of eleventh-century France, this was Halley's 1066 appearance, recorded in the Bayeux Tapestry.

PLUTO

The Chill-Out Zone

The outer limits

For long-haul travellers, the edge of the Solar System has many enticing destinations, although you'll need a suspended animation chamber for the long voyages between these tiny worlds. This is the realm of **Pluto**, **Eris** and a thousand worlds like it in the Kuiper Belt. Out beyond this belt lies the heliopause, where the stream of particles from the Sun meets the wind from the stars. Then comes Sedna, a mysterious red world that astronomers still don't understand. At the very limit of the Sun's gravity lies a spherical halo of countless dormant comets, the Oort Cloud, a quarter of the way to the nearest stars.

It's hard to believe, but some people have already become jaded with the wonders that space tourism has to offer. That's snobbery for you, we suppose. Who was it who once said 'He is a tourist, you are a holidaymaker, but I... am a traveller'?

So if you're the type of person that thinks the Moon's overcrowded, Mars is too common and even the rings of Saturn are going downhill fast, what's left? Well, there's plenty of room at the edges – myriad unexplored worlds just waiting to be discovered, named and mapped. The only catches are it'll take you the rest of your life to get there, and you won't be able to bring back fascinating native craft-works with which to impress your equally snobbish friends. In fact, you may not make it back yourself!

We're being unfair, of course. Not everyone who heads off to explore the outer edge of the Solar System is a tourism snob, and to be honest, only a few dozen people have made it out this far. Apart from scientific expeditions, the majority of explorers out here are wealthy retirees who can afford to kit out a large ship with enough supplies to last for years.

The outer edge of the Solar System splits neatly into two parts. Just beyond Neptune lies the Kuiper Belt of 'ice dwarf' planets that includes Pluto, Eris, and a host of other worlds up to about the size of our Moon. Way further out, there's the Oort Cloud, a spherical halo of comets that surrounds the Solar System. To be honest, no one's ever made it out this far, but we're including it here for insurance – we don't want anyone demanding their money back in a century from now because their battered

Until the New Horizons spaceprobe made the first flyby of Pluto in 2015, Earth-based images like this one of Pluto and its satellites were all we could see of the Kuiper Belt.

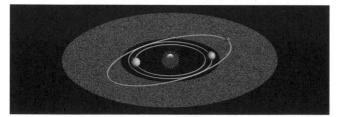

The Kuiper Belt extends from just beyond Neptune's orbit to more than twice that distance from the Sun. Pluto, long viewed as the ninth planet, orbits close to its inner edge.

copy of the *Traveller's Guide to the Solar System* didn't say anything about the Oort Cloud.

Getting there

Our main advice would be to get up as much speed as you can, and then go, go, go! That's trickier than it sounds, of course, especially when you're lugging several tons of supplies and life support equipment with you. If you like living dangerously and aren't too bothered about maybe losing a couple of toes to frostbite, you could save weight and money by taking a suspended animation unit with you. That way, you can sleep through the long boring bits between worlds (though we wish you luck getting insurance).

A slingshot around Jupiter is a must if you want to get there at any sane speed (see p.15), and trust us, the journey will be dull enough without adding another four or five years to it. Unless it's completely impossible, you'll want to schedule a few other flybys along your route. Alignments of all four major planets only happen every 176 years, but if you can fit in two or three, it'll help pass the time, and give you even more speed for the last leg of your journey.

INSIDE AN ICE DWARF

Ice dwarfs are all thought to have a pretty similar composition, and our illustrations shows Pluto, about which we understandably know the most. Despite the name, ice dwarfs aren't as icy as all that – Pluto is still about 70% rock, and the heat generated as it came together would have been enough to melt the planet, turn it spherical, and separate it into layers. The result is a rocky core, surrounded by a mantle of various ices dominated by water. The surfaces of the dwarfs seem to vary quite a lot, probably because they are formed from thin, multi-coloured frosts of various gases that can change from world to world.

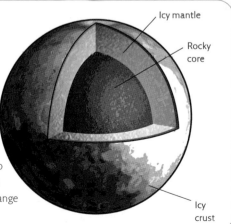

Icy mantle

Rocky core

Icy crust

Pluto

Going to the Kuiper Belt and not visiting Pluto is a bit like going to Egypt and not seeing the pyramids – it's pretty much unheard of, and you're not going to impress anyone with your lame excuses that it was on the wrong side of its orbit, you looked the other way and missed it, or your camera jammed.

Pluto was discovered way before any of the other Kuiper Belt Objects (KBOs), and spent most of the twentieth century as a bit of an embarrassment – a tiny world out on its own beyond the giant planets, messing up everybody's neat theories of how the Solar System formed. Some astronomers started saying 'It's not really a planet!' almost as soon as it was discovered, and it was finally bumped from the official planetary roster in 2006 (see p.199)

Time your visit to Pluto right, and you won't even have to travel as far as Neptune, since for 20 years of every 248-year orbit, Pluto dips inside Neptune's orbit. Unfortunately, it's almost impossible to visit both on the same trip, since Pluto's been nudged into an orbit that keeps it well away from Neptune around its perihelion. (This was a bit of a blow to the twentieth-century astronomers who hoped to explain Pluto away as a 'lost' satellite of Neptune).

During these close approaches, Pluto experiences a mild heatwave, with surface temperatures peaking around -230 °C (-382 °F). This is just enough to evaporate some of the ice patches on its surface, allowing Pluto to briefly develop an atmosphere. It's not much, but it's something: a thin blend of nitrogen, methane, and carbon monoxide. Pluto's day, night and seasons are complicated in the same way as those of Uranus, because its poles are also wildly tilted, knocked over at 122 degrees from the vertical. As a result, some parts of Pluto get a 100-year day, a brief period in which the sun rises and sets with the 6.4-Earth-day rotation period, and then a 100-year night. Night and day don't really make much

PLUTO DATA

Good points:
Nearby (for a Kuiper Belt Object)

Bad points:
Need to time your visit well to get the most out of it

Day length:
6.4 Earth days

Year length:
248 Earth years

Gravity:
0.06 g

Surface temperature:
-230 °C
-364 °F

Communications time:
4 hours or more

Approaching the Pluto system, you can look past the planet itself towards the major satellite Charon, and perhaps see one of the smaller moons too.

difference, though – this far out, the Sun's just a particularly bright star.

Pluto, like many other KBOs, isn't alone on its journey round the Sun: it has three moons for company, the largest of which, Charon, is more than half the diameter of Pluto. At 1,207 km (750 miles) across, Charon orbits incredibly close in – just 19,600 km (12,170 miles) away. Those ever-present tidal forces have slowed Charon's rotation so that it keeps one face permanently towards Pluto, but in this case they've gone one better, and slowed down Pluto so that it rotates in the same time that Charon takes to orbit – 6.4 Earth days. This makes the Pluto/Charon system a bona fide 'binary'. The two worlds keep one face permanently towards each other, and Charon never shifts in Pluto's skies.

The other two moons, Nix and Hydra, are quite a bit smaller, both about 50 km (30 miles) across. With orbits that keep them 49,000km (30,400 miles) and 65,000 km (40,400 miles) from Pluto respectively, they provide a great viewing platform for appreciating the central pair.

On the surface

One good piece of advice for explorers on Pluto would be to take a torch. Midday on Pluto is a like a moonlit night on Earth, and the icy surface can be treacherously slippery. The temperatures here are slightly warmer than Triton during perihelion, because the surface is generally darker and absorbs more of the Sun's feeble heat. However, Pluto has none of the ice geysers you'd see on Triton. The surface is ancient and cratered, and its excuse for an atmosphere forms by evaporating straight from the ice patches on the ground.

The most spectacular sight on Pluto is Charon. Land in the right hemisphere and it's unmissable, seven times the apparent size of the Moon as seen from Earth, and shinier and icier than the surface around you. It's a weird feeling, standing on the surface with this great big threatening ball of grey-brown rock and ice hanging above you, never moving.

PLANETS, SCHMANETS

The discovery of the Kuiper Belt opened a huge can of astronomical worms about how you define a planet. It's the kind of subject that only professional astronomers and museum curators get worked up about, but it seems that every few years someone decided to get themselves in the news by declaring that Pluto wasn't a planet, or announcing that they'd just found the 'real' tenth, eleventh, or nineteenth planet. To be honest, though, even before they started discovering KBOs bigger and better than Pluto, the list of planets was a bit of a mess – and most attempts at a scientific definition ended up either missing out some of the traditional nine, or including some of the asteroids and other minor bodies. The discovery that Pluto wasn't even the largest world of its kind just added to the fun (see over).

Finding Pluto

The tale of Pluto's discovery is 1% inspiration, 2% coincidence, and 97% sheer bloody-mindedness. The inspiration came from Percival Lowell (1855–1916), an American astronomer who had produced some of the best maps of the Martian canals (see p.83) – hardly the greatest recommendation, with hindsight. In the late nineteenth century, a lot of astronomers clearly felt that there hadn't been enough planets discovered lately, so they set about inventing new ones. In fairness, the inventions were a response to what seemed to be deviations in the orbit of Neptune, but some people ran further with the speculative ball than others. In the early 1900s, Lowell proposed a new planet of his own, giving it the snappy title 'Planet X'. He built a new observatory at Flagstaff, Arizona, just to look for it, but died before the new world could be found.

In 1929, 13 years after Lowell's death, a young astronomer called Clyde Tombaugh took on the task of searching for Planet X. To do this, he planned to photograph a broad region of sky around the ecliptic, snapping each area twice, several nights apart. That way, he could compare the two plates, and see whether anything had moved. Fortunately he had a special viewing machine called a 'blink comparator' to help highlight changes between the plates, but it was still a tedious task.

Amazingly, though, Tombaugh got lucky. He found Pluto within a few months, very close to Lowell's predicted position. Looking back, Tombaugh's luck was even better than it seemed at the time: Pluto turned out to be far too small to disrupt Neptune's orbit, and eventually Neptune's 'wobbles' were explained away with no need for another planet at all!

Just as Charon never shifts in Pluto's sky, so Pluto hangs perpetually over Charon's horizon.

Eris

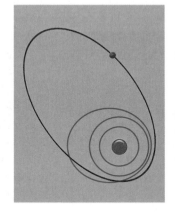

Eris's surface is surprisingly bright and grey for an object this far out in the Solar System.

Around the end of the twentieth century, Pluto lost its loner status in the outer Solar System, as its companions in the Kuiper Belt finally emerged. They'd previously been missed for a variety of reasons (Clyde Tombaugh, for one, carried on searching for another thirteen years without any luck). The drought was finally broken with the arrival of more powerful telescopes, and computers that could do much of the tedious image comparison work while astronomers hung around the coffee machine thinking up new research proposals.

Eris wasn't the first of the new KBOs to be found, but it was the first one bigger than Pluto. It was immediately trumpeted in the press as the 'tenth planet', but the International Astronomical Union (astronomy's top organisation) now had a headache. Official catalogues listed it as 2003 UB313, and the discoverers had nicknamed the new world Xena, after a popular Warrior Princess of the time, but before it could get a slightly less flippant official name, the IAU had to decide exactly what it was. To do this, they realised they'd finally need a decent scientific definition for a planet.

This map shows Eris's orbit in comparison to those of Pluto, Neptune, and Uranus.

So the IAU's general assembly got together in Prague, presumably hoping that the local beer would help ease a decision. In a heated week of debate, they toyed with one proposal that might have added dozens more planets to the list, before deciding that was a bit silly, and throwing it out. In the end, they opted for the more sensible choice of declaring that there were simply eight major planets: anything smaller was a dwarf planet, and that was that. This left a lot of people feeling sorry for Pluto, but at least Xena could finally be christened. Rather fittingly, they named it after the Greek goddess of discord and chaos.

And what of Eris itself? Well it's not exactly the most accessible object from the main Kuiper Belt, but this far out in the Solar System, orbital speeds are so slow that it won't take much of an engine burn to change direction. Eris is part of the 'scattered disc', a group of KBOs that were flung into unusually tilted orbits by close encounters with Neptune. As a result, it now ranges in distance from the Sun between 5.7 billion km (3.5 billion miles), about as close as Pluto at aphelion, and 10.2 billion km (6.3 billion miles). Its orbit is also tilted at 44 degrees to the plane of the Solar System. If you're heading this far out, you'd better have a working suspended animation unit and no one waiting for you back home. If not, we don't blame you for turning back, since no one else has ever got this far...

Having admitted that, we've only a sketchy idea about what you might expect of Eris. We do know that it takes 557 years to go round the Sun, and is only slightly larger than Pluto at around 2,397 km (1,489 miles) across. It's also extremely bright, more reflective than any other world in the Solar System except Saturn's snow-coated moon Enceladus. The surface is covered in a variety of ices, including methane, which might form a reflective frost. One other thing to look out for will be Eris's small satellite, originally nicknamed 'Gabrielle' after Xena's trusty sidekick, but now stuck with the rather less catchy name of Dysnomia. It's about 300 km (190 miles) across, and orbits Eris in roughly 14 days.

ERIS DATA

Good points:
Relatively easy to spot.
Nearby at times

Bad points:
Identity crisis.
A long way to go if
you time it wrong

Day length:
14 Earth days?

Year length:
557 Earth years

Gravity:
0.07 g

**Surface
temperature:**
-240 °C
-400 °F

**Communications
time:**
4 to 8 hours

Further afield — Sedna

The most distant individual world in our guide, Sedna is a loner even by the standards of the outer Solar System. It loiters awkwardly between the Kuiper Belt and the Oort Cloud, not wanting to join either party, but apparently content to spoil things for astronomers who like things neat and tidy.

Sedna was discovered in 2003 as part of a planned search for KBOs, but it turned out to be so slow-moving that it had to be much further away than the Kuiper Belt. When the discoverers worked out its orbit, they found that it takes 10,500 years to orbit the Sun, never coming closer than 11.4 billion km (7.1 billion miles), and getting out to as far as 146 billion km (91 billion miles) at its most distant. Little wonder, then, that they named their find after the Inuit ocean goddess, associated with the frozen seas of the Arctic.

Sedna just clips the outer edge of the Kuiper Belt at around 12 billion km (7.4 billion miles) from the Sun, but doesn't get as far out as the inner edge of the Oort Cloud (thought to be around 4,800 billion km (3,000 billion miles)

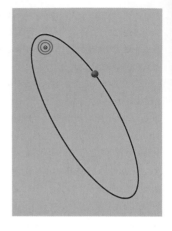

Sedna's orbit is so huge that even the outer planets are lost in this map – all that can be seen is the general zone of the Kuiper Belt.

Sedna lurks in the darkness of the outer Solar System, distant sunlight catching its ruddy surface.

away). Some astronomers suggested that it was evidence for an 'Inner Oort Cloud', but until someone finds some friends for Sedna following a similar kind of orbit, that will have to remain a theory.

So how much do we know about it, since nobody's gone there yet? Well, it's thought to be about 1,700 km (1,050 miles) across, it's pretty bright and so probably icy on the surface, and it's very, very red – about as red as Mars, to everyone's surprise. To save admitting that we just don't know why, we'll suggest that it could be something to do with organic chemicals on the surface.

One other thing: Sedna probably doesn't have a large moon. Tracking one down was an early priority after the discovery, because if one existed, it could be used to learn a lot more about this strange little world. Observers soon spotted changes to Sedna's brightness that suggested it rotates in about 20 days, and they thought this was a sign that Sedna's spin had been slowed by interaction with a moon (just as happened to Pluto). But despite several searches, no moon so far!

CROSSING THE HELIOPAUSE

Somewhere out beyond the Kuiper Belt, you'll take your first step into interstellar space as you cross the heliopause, the boundary where the solar wind smacks head-on into the oncoming galaxy. All the way across the Solar System, charged particles from the Sun have been keeping you company, racing ahead but gradually losing speed. At the 'termination shock', the particles slow to below the speed of sound. You'll then be travelling through the heliosheath, a roughly egg-shaped region of space dominated by the Sun, with the blunt end pointed in the direction of the Sun's motion through the Milky Way. Finally, at the edge of this, the solar wind has weakened so much that it's overwhelmed by the oncoming pressure of the interstellar wind – the drift of particles through interstellar space. As you cross the ensuing shockwave, some astronomers would say you're leaving the Solar System. But in our view, there's one final call to make...

SEDNA DATA

Good points:
A unique destination. Colourful

Bad points:
A long, long way to travel for not very much!

Day length:
20 Earth days?

Year length:
10.500 Earth years

Gravity:
0.05 g?

Surface temperature:
-260 °C
-436 °F

Communications time:
From 10 hours to 5.5 Earth days

Further afield —
The Oort Cloud

The Oort Cloud is the ultimate in hard-to-reach destinations, so far out that if you head in the right direction, you'll be a halfway to the nearest star before you cross it. It's a spherical shell with an inner edge about half a light year from the Sun (roughly a thousand times the distance of Pluto), and an outer edge around 1.5 light years away. Literally trillions of comets orbit here, but the space they occupy is so vast that they are even more thinly spread than the main belt asteroids.

This is the Solar System's backstage area, where dormant comets wait for their moment in the limelight. Most of them were born closer to the Sun, in the space where the giant

PREDICTING THE CLOUD

Considering the Oort Cloud is so distant and its contents so small, how can we be sure it's there? It's all a matter of statistics, and the Oort Cloud is named after the twentieth century Dutch astronomer Jan Oort, who was the first to work it out. Oort realised that long-period comets came from all directions, and their orbits could be inclined at any angle to the ecliptic, but their aphelion points always turned out to be in the same spherical region of space. In order to explain this distribution of comets, the shell-like Oort Cloud must exist.

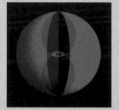

planets orbit now, but they were flung into the darkness by early encounters with the playground bullies, Jupiter and Saturn. Now they skulk around in million-year orbits, only occasionally encountering relatives in the void.

But every now and then, something stirs out in the Oort Cloud. Perhaps two sleepy nuclei brush close to one another, forever altering their orbits. Perhaps another star draws close, raising tides among the comets. Or perhaps the Solar System's passage through a particularly crowded region of space pulls them this way and that.

Whatever the cause, the result is the same – comets are shaken from their orbits and sent plunging towards the inner Solar System, gaining speed as they go. Thousands of years later, they may return here bearing the scars of their passage round the Sun. But many do not make it back. They may suffer another bruising encounter with Jupiter, which might destroy them completely, or pull them into the much closer orbit of a short-period comet. They may disintegrate with the stress of their passage close to the Sun, or they may smash headlong into a planet or moon. It may seem unlikely, but dozens of comets fall sunward each year, so it's more or less inevitable in a long enough time.

This is the true edge of the Solar System – the very limit of our Sun's gravitational grasp. If you don't go home now, you're probably never going to.

COMET CYCLES?

Every few million years or so, the Oort Cloud suffers a bigger disruption than usual, sending hundreds or even thousands of comets plunging towards the inner Solar System, and massively increasing the risk of a direct hit on a planet. But can we predict these cometary bombardments in advance? Some astronomers think so – they suspect there might be a regular period of around 26 million years in between bombardments, perhaps tied to waves of extinctions on Earth. If so, one cause might be the Sun's regular passages through the crowded plane of the galaxy. Another great idea, but one with no evidence to back it up, is that the Sun has a dim dwarf star companion orbiting somewhere in the wastes of the outer Solar System. At one point in its orbit, the theory goes, this 'death star' clips the inside edge of the cloud, sending a new wave of deadly comets towards the Sun.

A chance encounter between two dormant comets in the Oort Cloud may send one or both of them on a path towards the Sun.

Traveller's reference

No matter how well prepared you are, the Solar System can sometimes be a bewildering place. So this section of the guide offers you some background information on the history of our planetary system and its exploration, as well as discussing a few specific aspects of space travel, such as your choice of transport, food, water, and accommodation.

Historical background

A little understanding of the basic history of the Solar System never goes amiss. It all makes much more sense if you understand who did what to whom, where, when, and what with. It's a bit like an interplanetary game of Cluedo, only with asteroids instead of candlesticks.

The whole thing kicked off around 4.6 billion years ago. Our Solar System was little more than a collapsing ball of gas and dust, growing denser in the middle where the Sun would eventually form. As different regions orbiting in different directions collided with each other, their random motions were gradually flattened out into a disk around the Sun's equator. This 'protoplanetary nebula' must have looked a bit like Saturn's rings on steroids.

Stars are born in clusters, within towering, finger-like pillars of gas and dust such as these ones in the Eagle Nebula, 7,000 light years away from Earth.

Luckily for our future existence, the heavier particles and molecules in the nebula were quite sticky, so they tended to naturally clump together. Even before the Sun had begun to shine properly, asteroid-sized 'planetesimals' had formed in the disk around it. These baby worlds might have been small, but they already had enough gravity to pull in more material from around them. This meant that they could grow a lot faster than an object that just relied on random collisions with other material in the nebula. As they swept up material from the disk around them, they also began to spin more rapidly, generally around an axis that kept them 'upright' relative to the rest of the nascent Solar System, and spinning anticlockwise when looked at from 'above'.

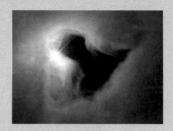

Individual stars form from relatively small knots of gas, perhaps a light year or so across, called Bok globules, which separate out of the larger nebula.

As nuclear fusion reactions finally kicked in at the heart of the Sun, the sheer force of radiation blasting out from its surface began to blow away much of the lightweight gas in the inner Solar System, and evaporate any easily melted ices in the area. Soon only rockier stuff remained, most of it bound up in a decreasing number of planetesimals zipping around in various eccentric orbits and eventually smashing into each other and merging together. The same forces that order planetary ring particles into flattened, concentric orbits also

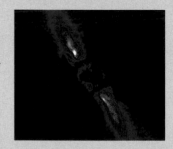

Infrared telescopes, which reveal material too cool to shine in visible light, have discovered many stars that are currently surrounded by planet-forming discs.

evened out the orbits of the planetesimals, so that the largest ones developed orbits closer to perfect circles.

In the outer Solar System, the same sort of thing was going on, but here the Sun's radiation was too weak to clear out the enormous amounts of gas and frozen chemical ices (such as water) which orbited along with the rockier material. As a result, the planets that formed here were much larger – they probably came into being as huge swirling knots of gas separated from their surroundings and collapsed inwards like mini-Solar Systems, producing complex satellite families.

Much of the leftover material around the giant planets coalesced to form small icy worlds – the comets. But as the giants grew ever larger, their gravity began to fling many of the smaller worlds aside, sending them out of the Solar System altogether, or dooming them to orbit forever in the distant Oort Cloud. The somewhat larger 'ice dwarfs' that formed beyond Neptune were less affected, and remain roughly where they formed. The gravity of Jupiter may also have tugged at material orbiting to its sunward side, thwarting any chance of it pulling itself into a substantial world, and resulting in today's asteroid belt.

By around 4.56 billion years ago, the planets were more or less as we know them today, although they still orbited through a hailstorm of smaller bodies that constantly bombarded their surfaces. A few larger rogue planetesimals still moved on elliptical orbits that put them on collision courses with the larger planets. One such collision at around this time created Earth's Moon. Others continued for some time – one obliterated much of Mercury's rocky mantle, and another knocked Uranus off its axis, but we don't have accurate dates for these events. All we do know is that they probably happened before about 3.9 billion years ago – an era called the Late Heavy Bombardment, when the few remaining massive planetesimals were soaked up by the larger worlds, creating features such as the lunar impact basins. This was the last major incident in the history of the Solar System, and everything else is just a postscript.

The Sun begins to form at the centre of a collapsing gas cloud.

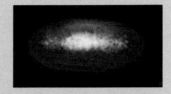

Material around the Sun collapses into a rotating disk, and begins to coalesce.

Close to the Sun, gas is blown away by radiation, and rocky planets form through collisions.

Further out in the Solar System, large whirlpools of gas collapse to form the giant planets and their moons.

The discovery of the Solar System

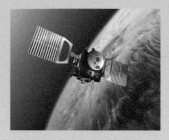

It's taken centuries to even begin to understand the worlds of the Solar System, but the first important steps were made by a series of robot spaceprobes in the late twentieth century – the first fifty years of what became known as the Space Age.

The Venus Express probe, above, was a sophisticated radar mapper that went into orbit around Venus in 2006.

The first few decades of space exploration were driven not so much by scientific exploration or commerce, as by a game of political one-upmanship between the United States and what used to be called the Soviet Union. The subtext for this 'space race' wasn't just, 'Aren't we great? We're so much smarter than the other guys!' There was also a heavy hint of, 'And watch out – we've got great big rockets too!'

One way of showing off just how big and powerful your rockets were, of course, was to launch things at targets further and further across the Solar System. So even before mankind had got the hang of launching satellites accurately and reliably, plans were afoot to invade other worlds.

Europe's Mars Express probe of the mid-2000s used three-dimensional photography to construct perspective views of features such as the Valles Marineris.

The first spaceprobes, naturally enough, targeted the Moon. Even with a big target just 400,000 kilometres (250,000 miles) away, though, they took a couple of tries to get it right. The Soviet-built Lunik 1 missed completely, and while Lunik 2 smashed into the lunar surface in 1959, it didn't send back any pictures. The first real success was Lunik 3, which beamed back the first pictures of the Moon's far side, also in 1959. The Soviets continued to send Lunik, Luna, and Lunokhod probes to the moon for the next couple of decades, but their thunder was soon stolen by America, which launched whole fleets of spacecraft at the Moon in order to survey it thoroughly before the manned Apollo landings of the late 1960s and 1970s.

Robotic rovers trundled across the Martian terrain early in the twenty-first century, searching for signs of water and life.

As rockets grew powerful enough to launch vehicles to the other planets, the Soviets concentrated on Venus, while the US had more luck with Mars. A series of Venera probes attempted to touch down on Venus, though this proved tougher than anyone had imagined. Later missions mapped the surface with radar. US Mars probes flew past the Red Planet several times, returning pictures of deserts and craters, before the first Mars orbiter, Mariner 9, revealed the far more interesting volcanoes, canyons and river valleys. NASA also managed to send the first probe to Mercury, using a gravitational slingshot (see p.15) for the first time.

By the time technology was advanced enough to explore the outer planets, the space race was almost over, and the Americans found they had the giant planets to themselves. After a couple of high-speed initial flybys of Jupiter and Saturn, they came up with a plan to send two probes on a 'Grand Tour' of the outer worlds, using a chance alignment between them to slingshot around each one and pick up speed for the next rendezvous. These were the legendary Voyager probes. Voyager 1 concentrated on Jupiter, Saturn, and Saturn's moon Titan, while Voyager 2 was able to fly on to the ice giants, and send back the first pictures of Uranus and Neptune. The initial trips to Jupiter and Saturn were soon followed up with the first probes into orbit around them – Galileo and Cassini respectively. Pluto, that perennial loose-end of the Solar System, was finally checked off the list of flybys by the New Horizons probe launched in 2006, still within the first fifty years of the Space Age (though it couldn't reach its target until the mid-2010s).

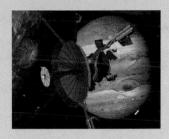

The Galileo spaceprobe spent six years in orbit around the giant planet, photographing its weather systems and moons.

The bus-sized Cassini probe orbited Saturn from 2004, revealing the surface of Titan for the first time.

SMALLER WORLDS

A whole host of early spaceprobes also investigated the smaller worlds of the Solar System, laying the foundations for our understanding of asteroids and comets. The first wave targeted Halley's Comet during its 1986 perihelion – it was too big and obvious to ignore. Galileo flew past a couple of asteroids on its way to Jupiter, and then NEAR spent a romantic year in orbit around Eros (see p.112). Several other probes investigated comets in various ways, collecting particles from their tails, attempting to land on them, and even smashing projectiles into them to see how they reacted!

Stardust was an early comet probe designed to sweep up material from the tail of a comet and return it to Earth.

Transport basics

One of the most important factors in any trip around the Solar System is your choice of transport. Before the Space Age, people had plenty of wacky ideas for getting around, but most of them would have ended up killing you in a variety of interesting ways. Then, for a long time, space travel was inextricably linked to just one means of transport – rockets.

A Chinese drawing of a typical gunpowder-fuelled military rocket from medieval times.

The first people to think seriously about getting into space were the great nineteenth-century writers H.G. Wells and Jules Verne. Both tried to be realistic in their depictions of space travel, but when it came to propulsion systems, they didn't really have a clue. In order to get *The First Men in the Moon* on their way to space, Wells invented a material called cavorite, capable of insulating a spacecraft from the effects of gravity. Verne was slightly closer to the mark, firing his spacecraft *From the Earth to the Moon* with a giant cannon. But he failed to realise that his heroes would have been squashed flat by the enormous acceleration of launch.

Ironically, the solution had been around for centuries. The Chinese had been using rockets in warfare and as fireworks since at least the thirteenth century, and the invention reached the rest of the world in the following 200 years. The two major problems with these rockets were that that they were hard to keep on course, and that traditional propellants were as weak as a kitten. This didn't stop nineteenth-century Russian revolutionary Nikolai Kibalchich planning the first man-carrying rocket, but it was a few more years before another Russian started to make the dream a reality.

Robert Goddard shows off his pioneering liquid-fuelled rocket in 1926.

We've met Konstantin Tsiolkovskii before – he's got a crater on the lunar farside named after him (see p.31). A schoolteacher in a provincial Russian town, he worked out a lot of the theory of rockets, explaining how they would be able to work in the vacuum of space (see box), how they could be steered, and how a multi-staged rocket with self-contained engine and fuel-tank stages would make it easier to reach orbit. Strictly an ideas man, Tsiolkovskii was

A captured V2 rocket at the US White Sands missile range shortly after World War II.

ignored at first, but after the Russian Revolution that founded the Soviet Union, his work was widely publicised.

It was an American, Robert Goddard, who took things to the next stage. He realised that a combination of liquid fuels and 'oxidants', such as hydrogen and oxygen would burn together and produce a heck of a lot more thrust than traditional solid fuels. This allowed him to launch the first liquid-fuelled rocket in 1926. It was also the vital remaining step in the quest to free rockets from the Earth: a traditional solid-fuel rocket still needed air around it as an oxidant so that it could burn its powder fuel. By carrying oxidant as well as fuel on board, a rocket could now fly anywhere.

Goddard and Tsiolkovskii's work excited a lot of young scientists in the 1920s and 1930s, and led to the creation of rocket societies in many countries. The most important of these was in Germany, where the 'Society for Space Travel', under the guidance of Hermann Oberth, developed new and powerful engines. Absorbed into the military machine by the Nazi party prior to World War II, society members such as Wernher von Braun (who later moved to the United States and masterminded the Apollo rockets) developed the first large, long-range rocket, the V2 missile. At the war's end, both the Soviet Union and the US scrambled to recruit German rocket scientists for their own missile and space programs. The result was a golden age of chemical rocketry.

Hermann Oberth (front), Wernher von Braun (centre right) and other members of the US ballistic missile development team.

A giant Saturn V rocket blasts free of the launch pad at Cape Canaveral, carrying the Apollo 11 astronauts on their way to the Moon.

ROCKET PRINCIPLES

The basic principle behind rocket propulsion is very simple. Rocket fuel and an oxidant are simply mixed together and ignited in a combustion chamber with an exhaust vent that points behind the rocket. As the fuel burns, the exhaust gases expand rapidly and are pushed out of the exhaust. Because, to quote Newton's laws, every action has an equal but opposite reaction, gas escaping in one direction pushes the rocket in the other. Before the Space Age, an awful lot of otherwise bright people didn't believe that a rocket could work in a vacuum. They thought the rocket got its forward thrust by somehow 'pushing' against the surrounding atmosphere.

Modern transport

Even today, you're going to be reliant on a chemical-fuelled rocket to get you off the Earth and into orbit. They might be noisy and uncomfortable, but they're pretty much the only type of vehicle that creates enough thrust to overcome gravity on Earth's surface. Once you get into orbit, though, you can choose between a range of interplanetary spacecraft, each of which has its advantages and disadvantages.

A nuclear ramjet offers a fast and fuel-efficient route to the planets, but that doesn't mean people have to like the idea.

Chemical rockets can still be used for interplanetary travel, of course, but they're comparatively inefficient for long-haul journeys, since you have to take a lot of fuel with you to reverse your acceleration and slow down again at your destination. Nevertheless, you're bound to use them at some points in the journey (they're still the standard for landing capsules, for example). If you're going to a planet with an atmosphere, you can save on fuel by aerobraking at the other end – dipping into the upper layers of the planet's atmosphere in order to slow down – but it's a hair-raising experience that we wouldn't recommend for first-timers.

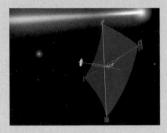

The acceleration available from a solar sail is almost limitless, but they can be impractical, and you'll need to keep your payload weight down.

Ion engines are an alternative that's reliable, quiet, and popular. There are various configurations, but they all work by splitting up atoms of a gas 'fuel' using electric sparks. The charged ions formed in this process are expelled from the engine's exhaust, pushing the spacecraft forward. One big advantage of this type of engine is that it can accelerate at a much slower, steadier rate, eventually reaching very high speeds. You can keep accelerating right up to the middle of your flight, then turn around and decelerate to your destination, generating artificial gravity all the while (see box). The major problem is that you need a reliable source of electrical power as well as your engine itself. Solar cells are effective near the Sun, but not so good in the outer Solar System, where the best option is a radioisotope thermal generator (a battery fuelled by the heat of decaying radioactive materials).

An ion engine is easy to spot from the distinctive blue exhaust glow. This is Deep Space One, a prototype space probe that tested the technology.

Talking of nuclear options, you could go for a completely atomic engine, generating thrust through a series of tiny nuclear explosions. The US space agency NASA first experimented with this idea in the 1950s, and it combines the higher thrust of a chemical engine with the sustained acceleration of an ion engine. However, most people aren't terribly happy at the thought of sharing their spacecraft with an atom bomb.

At the 'green' end of the spectrum, meanwhile, there are solar sails. These are perhaps the ultimate in fuel-efficient propulsion, since you don't really need fuel at all. The basic idea is that you unfurl several square kilometres of reflective silver foil in orbit, and that as light from the Sun bounces off it, it transfers enough momentum to push you forwards. The good points are that your energy supply never runs out and, because radiation from the Sun can always outpace you, you can reach very high speeds. The bad points are that it takes a painfully long time to get up speed, and that, no matter how much you use nautical techniques like tacking and jibbing, it's impossible to sail straight back into the solar wind, so in practice you'll need a fuel supply for your return journey.

A wacky alternative to solar sails is laser propulsion. The basic principle is the same, only instead of using light from the Sun, you're pushed forward by radiation from a laser beam fired at a reflector on the back of your ship. This avoids the need for a huge sail, but the boffins who came up with this one haven't really got the hang of keeping the laser on target, and you still have the problem of getting back!

Laser propulsion is a nice concept, but to make it practical we'd need to set up powerful laser beam generators across the Solar System, as shown in this artist's impression.

Artificial gravity (see below) isn't just good for your health. It also helps keep flyaway hair under control!

Artificial gravity

If your spacecraft can generate some form of artificial gravity, it will help to avoid a whole host of day-to-day inconveniences (for instance, dealing with food and water), not to mention health problems. The most direct way of generating the effect of gravity is simply to spin your spacecraft on its axis. This will give everything a natural tendency to drift towards the walls, but you'll use up quite a bit of energy keeping it spinning over a long trip. The sensible alternative is to use a propulsion system that can generate thrust throughout long periods of your flight (like an ion engine, but unlike a chemical rocket). That way, the constant acceleration or deceleration always pushes you in one direction or another, creating an up and a down. It's a little like taking the enormous g-forces of a rocket launch and spreading them out through your entire journey. The only problem is that, when you reach the midway point, turn around and begin to decelerate to your destination, 'up' and 'down' will reverse, so the floor will turn into the ceiling and vice versa!

Food and drink

Food for space is not the big problem that it once was. Back in the early days of space flight, doctors on the ground were worried that human digestion might not work without gravity to help it along, while the engineering guys were terrified of crumbs getting into their delicate wiring.

Fortunately, both fears proved exaggerated. As we mentioned in 'Up, up and away!' (pp,20–27), a bit of nausea is almost inevitable when you go weightless for the first time, but the human digestive tract relies more on muscle contractions than on gravity, and so it keeps working perfectly well. And just because you're eating in a zero-g environment, it doesn't automatically mean that your food is going to fragment into tiny pieces and go all over the place.

The real trick, though, is preparing the food in the first place, since that inevitably involves heating it up and, if it's been freeze-dried for long-duration storage, rehydrating it. Space ovens are straightforward enough – often they'll have a sort of 'hotplate' where you clip the foot container to make sure that it's thoroughly and efficiently heated up. Some meals are easy enough to eat straight from the container (a sort of zero-gravity TV dinner) and you can use the same clipping system to anchor your meal to a table if you prefer more formal dining. Other food requires a little more effort: you'll probably come across some dried foods prepacked in sachets with a nozzle at each end, or in concertina-style plastic containers. In both cases, you just add water through one valve (usually by inserting a tube from the onboard water boiler), and squeeze and expand the container until the food mix inside is thoroughly rehydrated. Then open the other valve and squeeze the paste you've just made out into your mouth. It's not the most elegant way to eat, but this isn't tea at the Ritz.

Although a lot of space food bears a strong resemblance to the worst kinds of ready meal you'd find back home, you needn't worry too much about your nutrition, since space

Looks delicious, doesn't it? This is a typical ready meal as eaten by astronauts of the early Space Age.

Lunch in zero-g is the ultimate in 'food on the go', but stopping to eat properly is still going to be better for your digestion.

Enjoy fresh fruit and vegetables while they last! Fortunately vitamin supplements will protect you from the dietary diseases that plagued the sailors of earlier generations.

meals are carefully designed to give you just the right balance of vitamins, minerals and food groups for daily life in orbit. Hopefully that'll be some comfort when you're halfway across the asteroid belt and pining for a home-cooked lasagne and freshly-tossed salad...

But can I drink the water?

We've already touched on the problems of dealing with fluids in space (see p.25). The golden rule is to keep any liquids safely contained whenever possible. However, water supply is a sensitive subject, and one that scheduled tour operators often prefer to avoid. This is because, like it or not, any long-duration space flight is going to involve a certain amount of recycling. Water's simply too precious a resource to allow it to go to waste: you can't compress it for storage, and every thousand litres you take with you will add another tonne to your payload. So the water vapour that you breathe out will be caught by filters in the air conditioning and sent back to the main tanks. The same goes for (how can we put this nicely?) other sources of water from your body. It's nothing to get worried about, since the purification systems are very thorough, and the residents of some Earth cities have been putting up with this sort of thing for ages. The real problem is just that, in the cramped confines of a spacecraft, it's all so much more obvious.

There are other sources of water out there, of course, and hopefully before too long they'll offer some respite from the recycling. Mercury, the Moon and Mars all have relatively pure water ice at their poles, while the moons of the giant planets have huge amounts of ice mixed with various amounts of rock. It's easy enough to extract the water from a mix of rock and ice by melting it, but you'll still need a portable desalination and purification plant to get rid of the various chemical salts that will be dissolved within. Nevertheless, it can only be a matter of time before some bright spark starts shipping Martian spring water back to Earth!

Eating a meal through a straw can be disconcerting at first, but you soon get used to it.

Just because water is at a premium, it's no excuse to stint on personal hygiene. A clean crew is a happy crew!

Although it's endlessly fascinating to watch a blob of water wobbling in zero gravity, you do need to watch out for vulnerable circuitry!

Accommodation

If there's one thing we can guarantee about the living conditions on your journey, it's that they'll be cramped. If you suffer from claustrophia, then the astronautical life probably won't be for you: outside of science fiction, no one's ever designed a spacecraft that could be described as light and airy. It's less of a problem for spacecraft assembled and launched from orbit than it is for those launched directly from Earth. Spaceplanes have to be aerodynamic to get through the atmosphere efficiently, while smaller capsules must at least be wrapped in a protective and aerodynamic cowling. Once you're in a vacuum, you could design a spacecraft in the shape of a brick and it wouldn't make much difference to its performance and handling. But extra space generally means extra weight, and as we've seen time and again, extra weight (or rather, mass) is a 'Bad Thing' for a spacecraft. Some early space station designers dabbled with the idea of inflatable cylindrical segments that would blow up to create spacious living quarters, but the idea has yet to catch on: people tend to get uncomfortable at the thought of nothing but a thin membrane protecting them from the vacuum of space.

Your quarters may be cramped or even claustrophobic, but a little personal space is still a must for any long-duration spaceflight.

Life in space can rapidly become uncomfortable in ways you hadn't even considered. For example, with no atmosphere around you to moderate the effects of the Sun, parts of your vessel may be baking hot in the daylight, while areas hidden in shadow are freezing cold at the same time. Heat-transfer systems in your vessel's hull will reduce the amount of expansion and contraction caused by these changes, but they don't do much to help the crew.

You'll also have to get used to the ever-present sunlight of interplanetary space. Spinning your spacecraft will at least help to create a sense that the Sun is moving, but you won't be able to mimic a proper cycle of day and night, and even with the blinds down, you'll find the ever-present cabin lights will screw up your efforts to get regular sleep. For most people, the only thing to do is wear airline-style eyeshades

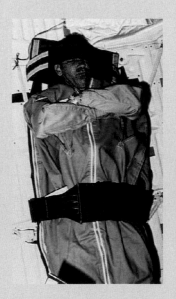

A lot of people find they sleep better when they're firmly secured to the floor (or wall, or ceiling). There's nothing more disconcerting than waking up adrift in weightlessness.

and try to ignore the incessant creaks, throbs, and whirrs of a spacecraft in flight. Securing yourself for sleep often helps, but it will probably take some time to get a new rhythm going.

One final word of warning: as you adjust to life in space, you'll probably feel like a bear with a sore head, but try not to take it out on your fellow travellers. There was a reason for all those psychological tests they put the early astronauts through, and even then, things occasionally went badly wrong. Try to remember that you can't take things back once they've been said, and there are few things worse than being trapped in a tin can millions of kilometres from home with someone that you loathe!

Ah yes – the toilet. Read the instructions *very* carefully before use! You don't have to worry about being sucked into the vacuum of space, but get things wrong in weightless conditions, and things could still get quite unpleasant for you and your crewmates.

MOVING ABROAD

People have always been tempted to sell up and make a permanent move to that lovely little spot where they spend their holidays, and it's almost inevitable that the same thing's going to happen in space, with civilian settlers moving out to join the scientific missions on other planets. Even today there are various organisations that are only too happy to flog you an acre of land on the Moon or Mars, but the reality is that Martian land deeds are unlikely to stand up in a court of law. The situation is more like the settling of the American Mid West (though fortunately this time there aren't any natives with a prior claim to the land...)

But life on the high frontier won't be glamorous. With transport costs so high, your home will probably turn out to be made from some junk that was going in the same direction anyway, such as a converted fuel tank. Even fitted out with a few home comforts, it's still going to be uncomfortably similar to a bomb shelter – especially considering the heavy layers of shielding you'll need on either the Moon or Mars to protect you from the solar wind.

Permanent colonies have to be within easy reach of a water source such as the polar caps of Mars or the ice craters at the lunar south pole. With water readily available, they'll still be reliant on food and other supplies from Earth, but they should be able to generate energy from nuclear fusion reactors, and even produce their own fuel supplies for short hops around the planet.

Mars settlements haven't quite grown to this scale yet, but this artist's impression gives a good idea of what a major settlement might look like. Note the Martian glider flying overhead.

Gardening will be a major pastime when our settlements grow larger and more self-sufficient, as suggested by another artist's impression.

Phrase book

Travelling around the Solar System, you're going to come across a lot of jargon. Hopefully this handy reference will make things simpler.

AMMONIA
A volatile chemical with the formula NH3 – meaning it's made of one nitrogen atom surrounded by three hydrogens. It's commonly found in frozen form among the outer planets, and helps to power cryovolcanism.

APHELION
The point in the orbit of an object orbiting the Sun where it is furthest from the Sun itself, and moving at its slowest speed.

APOGEE
The point in the Moon or a spacecraft's orbit around Earth where it is furthest from Earth and moving at its slowest. Also loosely applied to orbits of other planets.

ASTEROID
A rocky minor planet, typically found in the inner Solar System out to the orbit of Jupiter. The vast majority of asteroids are confined between Mars and Jupiter in the main Asteroid Belt, but there are plenty of escapees in other orbits.

ATMOSPHERE
A cloak of gas atoms and molecules held in place around a planet by its gravity.

AU
An abbreviation for astronomical unit, a common solar system measurement. An AU is the average distance from the Earth to the Sun – roughly 150 million km or 93 million miles.

CENTAUR
An icy minor planet in the outer Solar System, typically found from the orbit of Jupiter outwards. Centaurs are large, mostly dormant, comets.

COMET
A small icy object with a dark crust. Most comets inhabit the Oort Cloud, far from the Sun, but they are also found in and around the giant planets. A tiny proportion have been diverted into highly elliptical orbits that bring them close to the Sun. As they heat up around perihelion, gas and dust released by melting ice in the comet's 'nucleus' develops into an atmosphere or 'coma' and an extended 'tail' blowing away from the Sun.

CORE
The central and most compressed region of any planet or moon – usually the bit that retains heat longest.

CORIOLIS FORCE
A force around a planet generated by its rotation, and usually affecting the atmospheric circulation.

CRUST
The upper few kilometres of solid rock that forms the outermost layer of a terrestrial world like the Earth.

CRYOVOLCANISM
A special type of volcanic activity that happens on cold planets and moons, where a water-ammonia mixture acts like the molten magma in normal volcanism.

ECLIPTIC
The plane of the Solar System, passing through Earth's orbit and the Sun. Most of the other major and minor planets orbit on or close to the ecliptic.

EJECTA
The material thrown out during a meteorite impact, which falls back onto the landscape around the impact site and can create smaller secondary craters.

ELLIPSE
Technically an ellipse is a shape produced by cutting a plane across a cone. To the rest of us it's a stretched circle, with two 'foci' on either side of the centre. Orbits always follow elliptical shapes, and a circle is just a special kind of ellipse.

EVA

Extra-Vehicular Activity - the fancy name for a spacewalk.

Foci

Two points in an ellipse such as a planetary orbit, placed symmetrically around the centre of the ellipse. Their distance from the centre depends on the shape of the ellipse. The shared 'centre of mass' of the system, usually deep inside the object being orbited, sits at one focus.

Gravitation

A fundamental force of the Universe that causes all objects with mass to attract other objects towards them. According to Einstein, gravitation is actually caused by massive objects distorting the shape of space around them.

Gravitational Slingshot

An in-flight manoeuvre involving a close approach to a planet in order to pick up speed and change course.

Gravity

The force felt around any massive object that pulls other objects towards it or down onto its surface. The smallest objects with noticeable gravity are small asteroids and comets a couple of kilometres across.

Ice dwarf

A largish icy world orbiting the Sun in the Kuiper Belt beyond Neptune. The largest known ice dwarfs are Pluto and Eris.

Kuiper Belt

A doughnut-shaped ring of comets and ice dwarfs that rings the Solar System, beginning around the orbit of Neptune and extending roughly twice that distance from the Sun.

Leading hemisphere

The side of a planet or moon that faces along the direction of its orbit. For a moon in synchronous orbit, locked with one face permanently toward its parent planet, the leading hemisphere never changes.

Liquid metallic hydrogen

A form of hydrogen found inside the giant planets Jupiter and Saturn. Hydrogen atoms normally like to pair up into molecules, but in the high-pressure interiors of the giant planets, they can be split apart into individual atoms which then act like a metal, conducting electricity and helping to generate magnetic fields.

Low Earth Orbit (LEO)

An orbit a couple of hundred kilometres or miles above Earth, where spacecraft can orbit after launch and before setting off into the Solar System.

Mantle

The middle layer in the construction of a planet, lying between the relatively thin outer crust and the central core.

Minor planet

Any object in orbit around the Sun that isn't a full-blown planet.

Momentum

A property of any object with mass, found when you multiply its mass by its velocity (speed in a particular direction). Isaac Newton figured out that you need to apply force to change an object's momentum, and that momentum is usually 'conserved' in any interaction between objects – the reason that 'every action has an equal and opposite reaction'.

Moon (with a small 'm')

Any natural object in orbit around a planet, whether major or minor. Many moons formed at the same time and from the same material as their parent planets, but some are rogue minor planets captured during close encounters with giant planets, and others, like our own Moon (with a capital 'M') have more complex origins.

Organic chemical

Chemist-speak for any compound based on carbon, usually mixed with other elements such as hydrogen, nitrogen and oxygen.

Oort Cloud

A huge spherical shell containing countless dormant comets that surrounds the Solar System at a distance of around one light year. The comets in the cloud originally formed much closer to the Sun, but were ejected to their current orbits by encounters with the giant planets.

Orbit

Any closed path that one object takes around another one under the influence of gravity. Orbits are always elliptical paths, with the object being orbited, or the 'centre of gravity' at one of the ellipse's 'foci'.

Parallax

An effect that makes objects at different distances appear to move relative to one another due to the observer's motion.

Perigee

The point in the Moon or a spacecraft's orbit around Earth where it is closest to Earth and moving fastest. The term is loosely applied to orbits around other planets because few people want to struggle with terms like perijove.

Perihelion

The point in the orbit of any body orbiting the Sun where it is closest to the Sun and moving at its top speed.

Planet

A large object in an independent orbit around the Sun or another star. The problem with this definition is that no one's decided precisely how large a world must be to qualify. Today there are officially eight planets: four terrestrials and four giants, but there used to be nine. Pluto was eventually excluded after the discovery of Eris confirmed that it was just another, slightly larger than average, ice dwarf.

Resonant orbit

An orbit (typically of a moon) that has a precise relationship with the orbit of another body. For example, a moon that completes an orbit in half or two-thirds the time of its outer neighbour. Resonant orbits can keep bodies safely separated, or bring them into frequent close encounters.

Rocket

Any engine that generates thrust through the law of action and reaction. In a typical chemical rocket, an explosion inside a 'combustion chamber' forces exhaust out through the back of the rocket. In order to keep the total momentum of rocket and fuel the same, the rocket is pushed forward with similar force.

Satellite

Any object, whether natural or artificial, in orbit around another. Natural satellites orbiting planets are also called moons.

Solar System

The entire region of space dominated by the Sun's gravity and solar wind, and everything within it – a spherical region of space around 2 light years in diameter.

Synchronous orbit

An orbit in which an object rotates in the same time as it takes to complete an orbit, and so keeps one face permanently toward the planet or star that it is orbiting.

Tectonics

A process seen only on large rocky planets (mostly on Earth) where the crust is split into separate plates that float on top of the mantle. This creates all manner of geological activity as the plates jostle past each other, collide, or separate.

Trailing hemisphere

The side of a planet or moon that faces away from the direction of motion. See Leading hemisphere.

Volcanism

The eruption of magma (molten rock) from beneath a world's surface through a crack in the crust. Magma on the surface is usually called lava.

Afterword

So there you go, that's the Solar System from the Earth to the edge: a holiday theme park two light years across, offering every form of entertainment, adventure and sudden death you can imagine, and probably quite a few you can't. Taking in every destination in this book would probably require a couple of lifetimes and financial reserves to make some countries envious, but that's why we told you to choose your destination carefully!

Of course we can't hope to be comprehensive, since even today we don't know everything about the Solar System. There are still innumerable minor bodies to be discovered in between and beyond the larger ones we know about, and we can't be entirely sure that there aren't any bigger surprises lurking in the cold dark wastes beyond the Kuiper Belt.

But we hope this book has at least covered the highlights of the known Solar System, and maybe whetted your appetite for travel. Whether you decide to take things further, and splash out on the trip of a lifetime, or simply remain an armchair traveller, good luck!

THE SMALL PRINT:

The author and publishers wish to make it clear that, while every attempt has been made to ensure scientific and technical accuracy within the bounds of current knowledge, we can accept no liability for any space traveller, present or future, who uses this book as anything more than a travel guide. Always read the spacecraft manual before leaving orbit. All warranties void if taken beyond Earth atmosphere. This does not affect your statutory rights.

Further reading

Your choice of reading matter on any long-duration spaceflight can make the difference between an enjoyable holiday and mind-numbing boredom. Why not try some of these classic titles?

Cosmos, by Carl Sagan
London and Sydney, 1981, Macmillan

The Conquest of Space, by Willy Ley, illustrated by Chesley Bonestell
New York, 1956, The Viking Press

The New Solar System, Edited by Beatty, Collins Peterson, Chaikin
Cambridge, 1999, Cambridge University Press

This New Ocean: The story of the first Space Age, by William E. Burrows
New York, 1998, Random House

Universe, by R. Dinwiddie *et al.*
London, 2005, Dorling Kindersley Ltd.

Websites

www.nasa.gov
General NASA website, with links to various missions

www. hubblesite.org
Official site of the Hubble Space Telescope

www.space.com
Daily space news updates

www.planetary.org
The website of the Planetary Society

Index

Acknowledgements

All images by Pikaia Imaging except those listed below. Every effort has been made to trace and credit the copyright holders, but in case of error, the publishers will be happy to amend credits in future editions. Pikaia Imaging would like to acknowledge the invaluable assistance of Matt Fairclough and all at Planetside Software for their help in the making of this book.

r = right, l = left, t = top, c = centre, b = bottom. 12 NASA. 16 tr NASA. 17 br NASA. 18 t nasa. 25 t NASA. 26 t NASA. 26 b ISS. 34 t nasa. 34 t NASA. 35 cr NASA. 36 bl NASA. 37 cr NASA. 38 br NASA/NSSDC. 40 t NASA. 48 t, br NASA/JPL. 54 tr, trc, trl NASA/JPL. 57 t NASA. 70 t NASA/SOHO. 70 br: Royal Swedish Academy of Sciences . 71 NASA/SOHO. 72 NASA/SOHO. 72 NASA/SOHO. 73 SST/Royal Swedish Academy of Sciences . 76 tr NASA/EIT. 84 b NASA/JPL. 86 br NASA/JPL. 89 tr NASA/JPL/Caltech Mars Polar Lander . 90 tl NASA/Greg Shirah. 91 tr, b NASA/JPL. 92 t NSSDC/NASA. 93 tl ESA/DLR/FU Berlin (Gerhard Neukum). 95 cr NASA/JPL. 96 tr NASA/JPL. 98 tr NASA/JPL. 101 t NASA/JPL. 111 tr NASA/JPL. 112 tr NASA/JPL. 117 t NASA/JPL. 124 tr, b NASA/JPL. 129 t NASA/JPL. 130 tl NASA/JPL. 136 tr NASA/JPL. 138 tr NASA/JPL/DLR. 145 tr NASA/JPL. 146-147 b NASA/JPL. 150 tr, bl NASA/JPL. 151 cr NASA/JPL. 152 tr NASA/JPL. 153 tr NASA/JPL. 154 tr, bl NASA/JPL. 157 br NASA/JPL. 160 tl NASA/JPL/ESA/University of Arizona. 162 tl NASA/JPL/SSI. 165 tr NASA/JPL. 170 tr NASA/JPL. 175 tr, bl GPL. 176 tl, cl NASA/JPL. 178 br NASA/JPL. 179 tr, cr, br NASA/JPL. 182 br NASA/JPL. 184 tr NASA/JPL. 185 tr NASA/JPL. 186 cr NASA/JPL. 188 tr NOAO. 189 tr ESA. 192 tr NASA, ESA. 204 tr NASA/ETI, crNASA/STSCI, lr ESO. 206 tr ESA, cr ESA, lr NASA/JPL. 209 tr ,cr lr NASA/JPL. 210 tr,cr, lr NASA, . 211 tr, lr NASA. 212 tr, cr, lr NASA, . 213 tr NASA/Pat Rawlings (SAIC), cr NASA/JSC. 214 lr NASA/MIX, cr, lr NASA/JSC. 216 tr, cr, lr NASA/JSC. 217 tr NASA/JSC, cr, lr NASA-GRC.